Neuroscience
Intelligence
Unit

Heat Shock Proteins in Neural Cells

Christiane Richter-Landsberg
Department of Biology, Molecular Neurobiology
University of Oldenburg
Oldenburg, Germany

Landes Bioscience / Eurekah.com
Austin, Texas
U.S.A.

Springer Science+Business Media
New York, New York
U.S.A.

HEAT SHOCK PROTEINS IN NEURAL CELLS

Neuroscience Intelligence Unit

Landes Bioscience / Eurekah.com
Springer Science+Business Media, LLC

ISBN: 0-387-39952-6 Printed on acid-free paper.

Springer Science+Business Media, LLC, 233 Spring Street, New York, New York 10013, U.S.A.
http://www.springer.com

Please address all inquiries to the Publishers:
Landes Bioscience / Eurekah.com, 1002 West Avenue, 2nd Floor, Austin, Texas 78701, U.S.A.
Phone: 512/ 637 6050; FAX: 512/ 637 6079
http://www.eurekah.com
http://www.landesbioscience.com

Printed in the United States of America.

9 8 7 6 5 4 3 2 1

Library of Congress Cataloging-in-Publication Data

Heat shock proteins in neural cells / [edited by] Christiane Richter-Landsberg.
p. ; cm. -- (Neuroscience intelligence unit)
ISBN 0-387-39952-6
1. Heat shock proteins. 2. Nerve tissue proteins. 3. Neurons. 4. Neuroglia. I. Richter-Landsberg, Christiane. II. Series: Neuroscience intelligence unit (Unnumbered)
[DNLM: 1. Heat-Shock Proteins. 2. Neurons--physiology. 3. Neuroglia--physiology. WL 102.5 H437 2006]
QP552.H43H435 2006
573.8'5--dc22

2006025298

About the Editor...

CHRISTIANE RICHTER-LANDSBERG is a Professor of Molecular Neurobiology and Neurochemistry at the Biology Department of the Carl von Ossietzky University in Oldenburg, Germany. After graduating in Pharmacy at the University of Marburg, Germany, she received further academic degrees in Biology (Dr.rer.nat; Habilitation in Molecular Neurobiology) at the Universities of Göttingen and Bremen, Germany, and spent several years of postdoctoral time at the Weizmann Institute in Israel, at Harvard Medical School and Stanford Medical School in the United States, and during recent years as visiting Professor at Harvard Medical School and the University of Pennsylvania Medical School. She is a long-time member of the Society for Neuroscience, the International Society for Neurochemistry, the European Society for Neurochemistry and the German Neuroscience Association. Her research interests are the regulation of oligodendrocyte differentiation, death and survival, stress responses in nerve cells and glia, with a major focus on the oligodendroglia cytoskeleton during normal development and pathological situations.

To my mother Elisabeth, and my brother,
Thomas Landsberg, whose thoughts cannot fly any more

CONTENTS

EDITOR

Christiane Richter-Landsberg
Department of Biology, Molecular Neurobiology
University of Oldenburg
Oldenburg, Germany
Email: Christiane.Richter.Landsberg@Uni-Oldenburg.de
Chapters 1, 2

CONTRIBUTORS

André Patrick Arrigo
Laboratoire Stress Oxydant
Chaperons et Apoptose
Centre de Génétique Moléculaire
et Cellulaire
CNRS UMR 5534
Université Claude Bernard Lyon-1
Villeurbanne, France
Chapter 7

Luca Battistini
Neuroimmmunology Unit
European Center for Brain Research
Santa Lucia Foundation
Rome, Italy
Chapter 8

Celia F. Brosnan
Department of Pathology
Albert Einstein College of Medicine
Bronx, New York, U.S.A.
Email: Brosnan@aecom.yu.com
Chapter 8

D. Allan Butterfield
Department of Chemistry
Biochemistry and Molecular
Biology Section
University of Catania
Catania, Italy
Chapter 4

Vittorio Calabrese
Department of Chemistry
Biochemistry and Molecular
Biology Section
University of Catania
Catania, Italy
Email: calabres@mbox.unict.it
Chapter 4

Menotti Calvani
Department of Chemistry
Biochemistry and Molecular
Biology Section
University of Catania
Catania, Italy
Chapter 4

Maria E. Figueiredo-Pereira
Department of Biological Sciences
Hunter College of CUNY
New York, New York, U.S.A.
Email: pereira@genectr.hunter.cuny.edu
Chapter 6

Una FitzGerald
National Centre for Biomedical
Engineering Science
National University of Ireland
Galway, Ireland
Chapter 5

Anna Maria Giuffrida Stella
Department of Chemistry
Biochemistry and Molecular
Biology Section
University of Catania
Catania, Italy
Chapter 4

Olaf Goldbaum
Department of Biology
Molecular Neurobiology
University of Oldenburg
Oldenburg, Germany
Chapter 2

Adrienne Gorman
Department of Biochemistry
National University of Ireland
Galway, Ireland
Email: Adrienne.gorman@nuigalway.ie
Chapters 3, 5

Kate Reed Herbert
Neurobiology Group
Department of Biochemistry
National University of Ireland
Galway, Ireland
Chapter 3

Cesare Mancuso
Department of Chemistry
Biochemistry and Molecular
Biology Section
University of Catania
Catania, Italy
Chapter 4

Giovanni Pennisi
Department of Chemistry
Biochemistry and Molecular
Biology Section
University of Catania
Catania, Italy
Chapter 4

Sha-Ron Pierre
Department of Biological Sciences
Hunter College of CUNY
New York, New York, U.S.A.
Chapter 6

Afshin Samali
Neurobiology Group
Department of Biochemistry
and National Centre
for Biomedical Engineering Science
National University of Ireland
Galway, Ireland
Email: Afshin.samali@nuigalway.ie
Chapters 3, 5

Krzysztof Selmaj
Department of Neurology
Medical University of Lodz
Lodz, Poland
Chapter 8

Vita Vernace
Department of Biological Sciences
Hunter College of CUNY
New York, New York, U.S.A.
Chapter 6

Zhiyou Wang
Department of Biological Sciences
Hunter College of CUNY
New York, New York, U.S.A.
Chapter 6

Andreas Wyttenbach
Neuroscience Group
School of Biological Sciences
University of Southampton
Southampton, U.K.
Email: aw3@soton.ac.uk
Chapter 7

PREFACE

Heat shock proteins (HSPs), also called stress proteins, are not only induced in response to elevated temperatures, but also as a result of various stress situations, including environmental strains, viral infection, ischemia, anoxia and oxidative stress. These stress situations trigger cellular defence mechanisms that act as an emergency system capable of combatting the toxic consequences due to the accumulation of misfolded proteins.

Heat shock proteins are involved in many physiological processes, including development and differentiation, organisation of the cytoarchitecture by binding to cytoskeletal elements and regulation of the balance between cell death and survival. Many heat shock proteins work as molecular chaperones. In this role, they contribute to in vivo protein folding and prevent nonproductive interactions with other proteins and cellular components. In recent years it has been found that the chaperone system and the proteolytic machinery work closely together, and that proteasomal inhibition causes the upregulation of stress proteins. Impairment of the proteasomal machinery and chaperone functions lead to protein damage, which contributes to neurodegenerative disorders and to the aging process.

The brain is the most sensitive target for traumatic events. Many disorders of the nervous system are caused by intracellular or extracellular deposits of protein aggregates. Stress responses in the brain, as monitored by the induction of heat shock proteins, occur in distinct regions and affect nerve cells and glia in a cell-type-specific manner. Hence, stress proteins may serve as biomarkers and provide diagnostic tools that allow us to identify stress specificity and localize pathological processes. Their critical involvement during neurodegeneration in brain disorders, such as Alzheimer's, Parkinson's and Huntington's diseases, and multiple sclerosis, makes them promising candidates for therapeutic intervention and drug development.

The authors of this book address the role of heat shock proteins and cellular stress responses in the brain, with a major focus on nerve cells and glia, namely oligodendrocytes and astrocytes. Chapter 1 summarizes the classification and functional roles of heat shock proteins which are particularly abundant in brain cells. The next chapters deal with the physiological functions of heat shock proteins in the regulation and maintenance of the cytoskeleton (Chapter 2), their role in the regulation of neuronal differentiation and development (Chapter 3), and cell death (Chapters 4 and 5). Chapter 4 discusses the specific antioxidant capacity of the small stress protein HSP32 (the heme oxygenase-1), a heat shock protein with enzymatic functions. This chapter also highlights acetylcarnitine, a compound with neuroprotective benefits. Chapter 5 concentrates on the role of heat shock proteins in the regulation of programmed cell death in association with neurodegenerative disorders. In particular, the authors review their interference with apoptotic pathways and damage of the mitochondria, which play a key role during apoptosis. The following three chapters concentrate on heat shock proteins in neuropathological processes: Chapter 6 deals with the basic cellular role of the ubiquitin/proteasome pathway and its cooperation with the chaperone system, which is required to prevent the development of potentially toxic protein aggregates. Chapter 7 summarizes how stress proteins counteract common pathologies of "folding diseases" by reversing early synaptic and axonal abnormalities. Chapter 8 demonstrates the potential impact of certain families of heat shock proteins in multiple sclerosis, emphasizing their role in the generation of the immune response in more chronic phases of the disease process.

Research in the field of heat shock proteins is very active, and the authors of this book hope that their contributions will help neuroscientists bring together disparate data sets and develop new experimental strategies that will further elucidate the roles of HSPs in the nervous system during health and disease.

Christiane Richter-Landsberg

Acknowledgements

Life in a laboratory is full of excitement, accompanied by disappointment and frustration. I want to thank the many diploma and graduate students who have shared my enthusiasm in research, and over the years each contributed small pieces to the puzzle.

I am also grateful to Dr. Virginia Lee and Dr. John Trojanowski, at the Center for Neurodegenerative Disease Research, University of Pennsylvania Medical Center, U.S.A., for their encouragement, hospitality and many discussions. Finally, I thank Dr. Peter Richter, Benjamin Richter and Sebastian Richter for their support and understanding.

CHAPTER 1

Heat Shock Proteins:

Expression and Functional Roles in Nerve Cells and Glia

Christiane Richter-Landsberg*

Abstract

Heat shock proteins (HSPs) or so called stress proteins have multifunctional roles and are involved in many physiological processes, such as cell cycle control, cell proliferation, development, organisation of the cytoarchitecture, regulation of cell death and survival, and play regulatory roles in cellular aging and longevity. They participate in protein synthesis, protein folding, transport and translocalization processes, by acting as molecular chaperones. As a result of a variety of stress situations, HSPs accumulate and help to prevent protein misfolding and aggregation, provide tolerance against further stress situations, and cooperate with the ubiquitin proteasome system during protein quality control. HSPs are differentially expressed in nerve cells and glia, and cell type specific responses to various stressors are observed. Stress proteins may serve as biomarkers to identify stress specificity and localize pathological processes, leading to cell and organelle damage in the nervous system. They can be used as neuropathological markers and are promising targets for therapeutic intervention and drug development.

Introduction

Heat shock proteins (HSPs) also termed stress proteins are a large group of highly conserved proteins which are found in plant, bacterial, yeast and mammalian cells. During stressful situations, such as hyperthermia, ischemia, anoxia, viral infection, oxidative and environmental stress, HSP production is transiently increased providing cytoprotection and physiological defense against harmful cellular stress, which may confer tolerance against further stress situations. HSPs have multifunctional roles in eukaryotic cells, and under physiological and nonphysiological conditions are involved in maintenance of the cell cycle and regulation of cell proliferation, regulation of gene expression and developmental processes by interacting with signaling transduction molecules,[3,4] organisation of the cytoarchitecture by binding to and stabilization of cytoskeletal elements (Fig. 1).[5-8] Moreover, they play regulatory roles in cellular aging and longevity, and direct cells to apoptosis or necrosis.[9,10] Several of these issues will be addressed in the following chapters of this volume.

HSPs are classified according to their molecular weights, and are divided into five major families: HSP100, HSP90, HSP70, HSP60 and the small HSPs (sHSPs) including αB-crystallin with molecular weights ranging from 12 to 43 kDa.[11,12] Furthermore members of the glucose-regulated proteins (Grps) are markers for stress responses, they are induced by perturbation of the endoplasmatic reticulum.[13] One large group of HSPs, including HSP90, HSP70, HSP40 and sHSPs function as molecular chaperones and participate in protein synthesis, protein

*Christiane Richter-Landsberg—Department of Biology, Molecular Neurobiology, University of Oldenburg, POB 2503, D-26111 Oldenburg, Germany.
Email: Christiane.Richter.Landsberg@Uni-Oldenburg.de

Heat Shock Proteins in Neural Cells, edited by Christiane Richter-Landsberg.

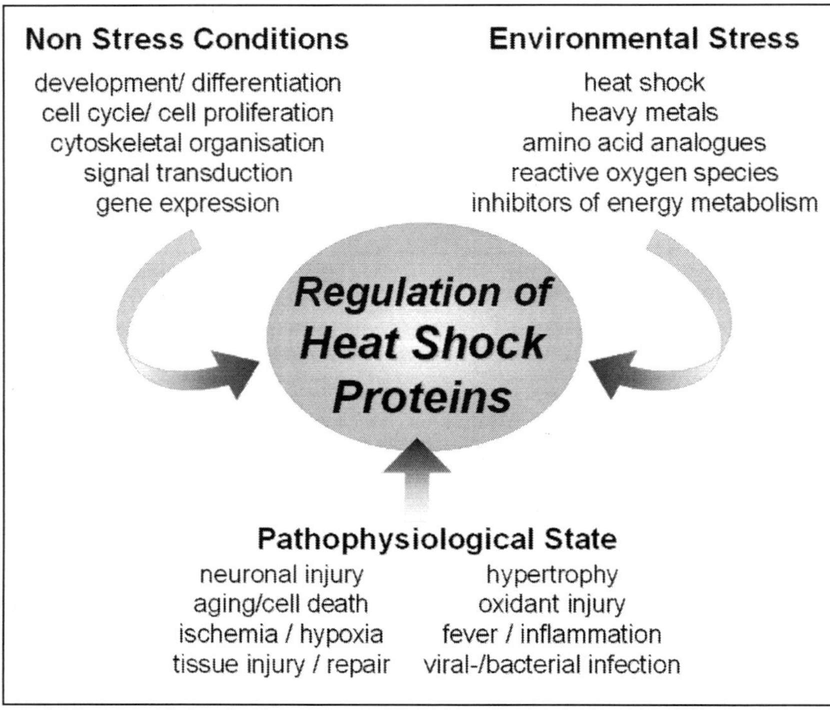

Figure 1. Heat shock proteins (HSPs) have multifunctional roles during health and disease. HSPs are modulated under physiological and nonphysiological conditions, and the expression of constitutive and inducible HSPs is regulated during cellular responses.

folding, transport and translocalization processes, and upon stress prevent irreversible aggregation of proteins (Fig. 2). Some of them show a broad specificity of binding to nonnative proteins, relatively stable complexes are formed and dissociated by the binding and/or hydrolysis of ATP.[11,14,15] Chaperones also prevent nonproductive interactions with other proteins and cellular components, such as the cytoskeleton and cell membranes. They bind to transiently exposed hydrophobic surfaces in proteins, shielding them from unwanted, aberrant interactions with other proteins, in an attempt to prevent the accumulation of aggregated or misfolded proteins. Overexpression of HSPs in a variety of yeast and mammalian cells was not successful in preventing protein aggregate formation, but reduced cytotoxicity and protected against degeneration. Hence, molecular chaperones possibly are important in modulating the aggregation process without necessarily inhibiting the formation of protein aggregates.

Furthermore, HSPs may promote the ubiquitination of abnormal proteins which are guided to and degraded by the proteasomal machinery (Fig. 2).[16] Malfunctions of the proteasomal system may cause serious pathological events and contribute to human diseases, including cancer and neurodegenerative disorders accompanied by amyloid deposits.[16,17] Molecular chaperones and the ubiquitin proteasome system (UPS) closely work together (Fig. 2), by cooperating during protein quality control and in determining the fate of proteins within the cells.[18] The molecular basis of this cooperation has not been fully elucidated, but it seems likely that chaperones, cochaperones and cofactors that associate with the chaperone/ substrate complex, such as CHIP (carboxyl terminus of HSP70 interacting protein), the BAG-family of proteins, hip (HSC70-interacting protein) or hop (HSP70/HSP90-organizing protein) modulate and determine the intracellular balance between protein folding and protein

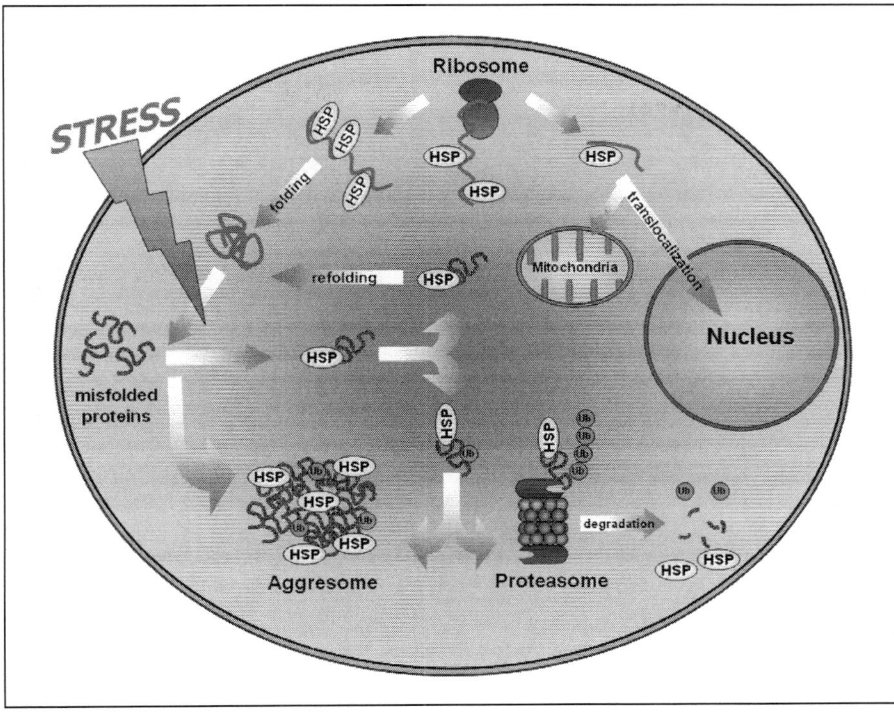

Figure 2. Illustration of the multiple roles of heat shock proteins in protein folding and during stress situations. Under physiological conditions HSPs bind to the nascent polypeptide chain emerging from the ribosome, and participate in their folding and assembly. They also are important in targeting proteins for translocalization to the nucleus and the mitochondria. Stress induces the formation of denatured proteins, which lead to the induction of HSPs. HSPs bind to denatured proteins in an attempt to refold them and to prevent their aggregation. If this is unsuccessful, HSPs may promote the ubiquitination of abnormal proteins which are guided to the proteasomal machinery (UPS) for degradation. If the capacity of the cells to degrade or refold abnormal polypeptides is exceeded, denatured proteins accumulate, tend to aggregate and aggresomes may be formed.

degradation.[18] Impairment of the proteasomal machinery by proteasome inhibitors, such as lactacystin or MG-132, cause the induction of HSPs and the accumulation of ubiquitinated proteins.[19-21] Hence, proteolytic stress is a potent inducer of HSP production. Also, among the genes induced upon heat shock are genes encoding ubiquitin and several ubiquitin-carrier proteins,[22] emphasizing that the two main protective strategies, protein degradation and repair of damaged proteins, are closely linked processes.

The heat shock response and rapid production of HSPs is regulated by heat shock transcription factors (HSFs) (Fig. 3). Multiple HSFs have been identified in vertebrates and plants, indicating specialized functions of the different HSFs during physiological and environmental stress. In mammalian cells, three HSFs, namely HSF1, HSF2, and HSF4, have been identified.[23] HSF3 is expressed in avian cells. In vertebrates, HSF1 is the major regulator of the heat shock response and other environmental stressors. Under nonstressed conditions HSF1 is in an inactive monomeric form, partially phosphorylated, with little DNA-binding activity. Upon stress, HSF1 trimerizes, acquires DNA-binding activity and becomes hyperphosphorylated. HSF1 regulates heat shock gene expression by binding to highly conserved heat shock elements (HSEs) in the promoter regions of heat shock genes (Fig. 3).[24,25] Phosphorylation plays a

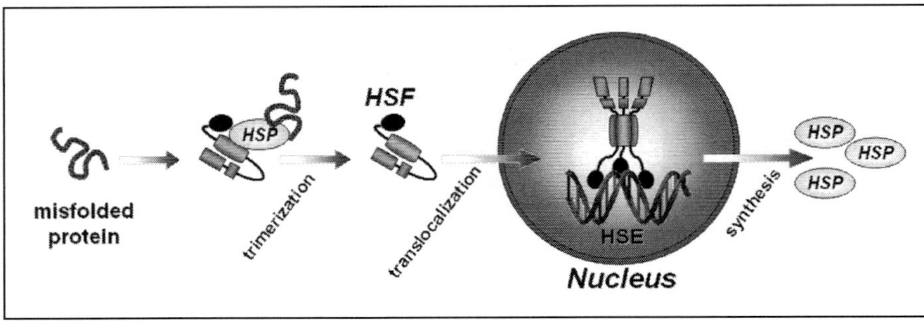

Figure 3. The heat shock response and rapid production of HSPs is regulated by heat shock transcription factor (HSF1). Upon stress, inactive HSF1, binding to HSPs, is released, followed by trimerization, phosphorylation and translocation to the nucleus. HSF1 binds to the heat shock element (HSE) in the promoter region of the heat shock genes, and induces the synthesis of HSPs. This response is then downregulated by high levels of chaperone activity which is sensed by HSFs. Particularly HSP40 and 70, which associate with the activation domain of HSF1, and also HSP90 have been shown to maintain HSF1 in an inert state.

major role in the regulation of transcriptional activation, but seems to be not important for the acquisition of HSF1-DNA-binding activity. HSF1 activity is negatively regulated by the chaperones HSP70 and HSP40, which associate with the activation domain of HSF1, and also HSP90 has been shown to play a role in maintaining HSF1 in an inert state.[26] Hence, the stress response is downregulated by high levels of chaperone activity which is sensed by HSFs. Proteasome inhibition has been shown to activate all three members of HSFs in avian cells,[27] and the proteasome inhibitors lactacystin and MG-132 induced the hyperphosphorylation and trimerization of HSF1 and upregulate HSP gene transcription.[28] Another study indicates that the heat shock response upon proteasomal inhibition seems to be specifically dependent on HSF1 as the critical transcription factor.[29]

Stress-responses in the brain as monitored by the induction of HSPs occur region- and cell-type specific, and either nerve cells or glia are the primary targets. The molecular mechanisms underlying the cell-type specific responses are far from understood. In the following I briefly summarize the properties of HSPs expressed in nerve cells and glia, which are particularly relevant to the central nervous system (CNS).

Heat Shock Proteins in Nerve Cells and Glia

HSPs are located in the cytosol, and in eukaryotic cells they can be associated with various cellular compartments or are translocated upon stimulation, e.g., to mitochondria, ER or the nucleus. A number of heat shock proteins are constitutively expressed in the nervous system, and a variety of cell culture studies have been carried out on the presence and upregulation of HSPs. Among these are the HSP90 family, HSP70 family, HSP60, the small HSPs including αB-crystallin, and HSP32 and ubiquitin. Their upregulation has been connected to neuroprotection in neural cells.[30-32] Individual cells in the brain show different susceptibility to stress situations. Depending on the level of stress, some cells remain unaffected, while others are severely stressed and react by the induction of stress proteins. When a certain threshold is passed, the stress response can cause cellular dysfunction, and instead of playing a protective role, HSP induction possibly contributes to the onset of cell death (see Chapter 5 this volume). Some of the HSPs are differentially expressed or inducible in nerve cells and glia, e.g., astrocytes, oligodendrocytes and microglia. Oligodendrocytes, the myelin forming cells of the CNS, are specifically sensitive to stress situations,[21,33] and in this respect resemble nerve cells, while astrocytes are much more resistant. Filamentous inclusions containing HSP have been observed in all three cell types in a number of neurodegenerative diseases,[12,17] and elevated levels of HSPs have been identified during inflammatory responses, e.g., multiple sclerosis (MS).

HSP90, a Chaperone with Regulatory Functions in Signal Transduction

HSP90 is highly conserved and one of the most abundant proteins in eukaryotic cells, comprising 1-2% of the total soluble proteins in unstressed cells.[34,35] It is constitutively expressed in the mammalian brain[36] and present in cultured nerve cells, astrocytes, oligodendrocytes and microglia (Goldbaum and Richter-Landsberg, unpublished). Upon stress it is typically induced only a few-fold; a fraction thereof translocates to the nucleus, where it might be involved in signaling events of the apoptotic cascade.[10] Another member of this family is the glucose-regulated protein Grp94, which is located in the endoplasmatic reticulum (ER) and is induced in response to ER stress.[13] HSP90 has chaperone activity, but unlike other molecular chaperones, does not participate in general protein folding. HSP90 cooperates with HSP70 and other cochaperones in a multichaperone complex in the degradation of misfolded proteins.[37] Most of its known substrates are substrates involved in signal transduction and cell cycle control.[35,38] Hsp90 binds to these substrates, termed client proteins, and assists in final folding and stabilization, and regulates their half life. Upon stress, Hsp90 keeps nonnative client proteins in a folding-competent structure and prevents irreversible denaturation, a process which requires ATP-binding. When ADP is bound, Hsp90 promotes ubiquitination and degradation by directing client proteins to the proteasome.[39] Hence, Hsp90 is involved in the balance between protein stabilization and degradation of its client proteins.

Two major cytoplasmic isoforms of HSP90 exist, HSP90α represents the major inducible isoform and HSP90β is the constitutive minor isoform.[40] HSP90 is mainly a constitutive dimer (αα or ββ), the α-isoform, however, more readily dimerizes than the β-isoform. Recently a novel member of the HSP90 family, namely HSP90N, which represents a 75 kDa Raf-associated protein, was identified.[41] Functional differences between the isoforms in cell differentiation and development have been implicated,[40] but not much is known about their differential roles in the nervous system. HSP90 binds to cytoskeletal elements, such as actin and tubulin, and might be important in stabilizing the cytoskeleton and maintaining the cytoplasmic meshwork termed microtrabecular lattice.[7] HSP90 is involved in the regulation of cell death and survival by interacting with and stabilizing protein kinases.[42] In live cerebellar neurons HSP90 α and β was found to be localized at the cell surface where it might have a role in cell migration.[43] HSP90 may play an important role in synaptic transmission, both at the pre and postsynaptic compartment. Gerges et al[44] recently demonstrated that HSP90 is necessary for efficient release of neurotransmitter at the presynaptic terminal and a critical component in delivering AMPA receptors into the postsynaptic membrane. Also, retrograde transport of the glucocorticoid receptor in neurites has been shown to require the dynamic assembly of complexes with HSP90.[45]

Interesting perspectives have come from the development and use of drugs that are effective inhibitors of HSP90, which have mainly been used in their capacity as antitumor drugs. The first inhibitor was geldanamycin, a naturally occurring benzoquinone ansamycin isolated from *Streptomyces hygroscopicus* with antitumor effects.[46] Subsequently more stable geldanamycin analogues with less toxic effects were synthetically derived, e.g., 17-AAG (17-allylamino-17-demethoxygeldanamycin).[47,48] Geldanamycin directly binds to the N-terminal ATP-binding site of HSP90, and destabilizes the HSP90-multichaperone complex with its client protein. This results in stabilizing the conformation of the HSP-multichaperone complex, which directs HSP90 client proteins to the proteasome and ubiquitin-mediated degradation.[38,49] Thus, the hallmark of ansamycin activity is to induce the degradation of HSP90 client proteins by the ubiquitin-dependent proteasomal pathway.

In a cell culture model of Huntington's disease geldanamycin was shown to activate a heat shock response, i.e., HSP40, 70 and 90 were induced, and the aggregation of huntingtin protein was suppressed in a dose-dependent manner.[50] A recent study by McLean et al[51] indicates that geldanamycin reduced α-synuclein aggregation and prevented its cytotoxicity in human H4 neuroglioma cells, possibly by induction of HSP70. In this study the role of HSP90 was not evaluated. Geldanamycin was also effective in inducing HSPs in the brains of male rats and protecting against focal cerebral ischemia.[52]

Hence, these compounds might not only be promising as anticancer drugs, but also have potential in elucidating the functional roles of HSP90 in the nervous system, and are therapeutic candidates in eliminating or preventing protein aggregates which occur in neurodegenerative diseases.

HSP70 Family

The 70 kDa heat shock proteins include the constitutively expressed, only moderately inducible, cognate protein HSC70 with an apparent molecular weight of 73 kDa, its closely related inducible form HSP70, with an apparent molecular weight of 72 kDa, and the glucose-regulated protein Grp78 or BiP (for binding protein) which is located in the lumen of the ER. HSP70 are most abundant and have been extensively characterized after their initial discovery. All members of this family bind to ATP and comprise chaperone activity. HSP70 members recognize and bind to nascent polypeptide chains as well as to partially folded intermediates of proteins, preventing their aggregation and misfolding. The binding of ATP triggers a critical conformational change leading to the release of the bound substrate protein.[11] HSC70 participates in folding and assembly of nascent proteins, and maintains proteins in their unfolded state. Furthermore, it promotes protein translocation through intracellular membranes. HSP70 interacts with HSP40, and together they promote cellular protein folding and repair misfolded proteins.[53-55] HSP70 activity is assisted by cochaperones, such as BAG-1 or CHIP, (for example, see refs. 11,37).

It is generally accepted that HSP70 overexpression in neurons and glia leads to neuroprotection.[32] The constitutive isoform, HSC70, and its cooperator HSP40 are present in the postsynaptic structures of rat brain, where they might be related to synaptic plasticity phenomena.[56] While in nerve cells HSP70 is only weakly inducible,[57-60] it is highly upregulated by hyperthermia and a variety of other stress situations in astrocytes, oligodendrocytes and microglia, respectively.[33,61-67] HSP70 is involved in the regulation of cell death and survival,[10,68] and its expression has been linked to neurodegenerative and inflammatory disorders of the nervous system. These issues will be raised in some other chapters of this volume.

HSP60

HSP60 is a highly abundant protein and constitutively expressed in eukaryotes. HSP60 is synthesized in the cytoplasm and transported to the mitochondria, where it is associated with the mitochondrial matrix and participates in the folding and assembly of transported proteins into the mitochondrion.[69,70] Its presence in cellular compartments other than the mitochondria and also its location at the cell surface has been suggested.[71] The members of the HSP60 family, which are also called chaperonins, are ring-shaped oligomeric protein complexes which bind nonnative proteins within a large cavity.[11,72] Its functions are dependent on HSP10, which binds to HSP60 and regulates its substrate binding and ATPase activity. HSP60 is constitutively expressed in neuronal cells,[73,74] astrocytes[75] and oligodendrocytes.[8,12,76,77] In addition to its function in intracellular protein folding, HSP60 can stimulate cells to produce proinflammatory cytokines.[72] HSP60, similarly to HSP70, can elicit strong immunological reactions and is a target for autoimmune attacks,[70] and might play an important role in autoimmune diseases (see Chapter 8 by Brosnan et al, this volume). Furthermore, HSP60 may play a passive role in amplification of the caspase cascade and act as a positive regulator of apoptosis (see Chapter 5 by FitzGerald et al, this volume).[78]

The Small HSPs and αB-Crystallin

The group of small heat shock proteins (sHSPs) with molecular weights in the range of 12 to 43 kDa are closely related and includes the αA- and αB-crystallins.[79,80] They all contain a highly conserved carboxy-terminal region termed the α-crystallin domain.[11,81] Alpha-crystallins are the major structural proteins in the vertebrate eye lenses. The small HSPs tend to form large relatively unstable oligomeric complexes consisting of 9-50 subunits. These are formed from stable dimers that aggregate into tetramers.[82] The number of sHSPs and abundance varies considerably in different species.[80] In mammalian cells only one small HSP is expressed, HSP25 with a molecular weight of 25-28 kDa in rodents and its human homologue HSP27 with a molecular weight of 27-28 kDa.[70] HSP27 is regulated by phosphorylation through serine protein kinases, and phosphorylation is increased upon various stimuli, such as heat shock, sodium arsenite, oxidative stress, mitogens, tumor promoters and calcium ionophores (for examples, see ref. 83). Phosphorylation decreases the oligomeric size, which in the case of HSP25/27 is accompanied by a decrease in chaperone activity. Dimers of HSP25 do not exhibit chaperone activity, while large oligomers comprise high chaperone activity.[84] Large aggregates of α-crystallin (800kDa) consist of two types of related subunits: acidic αA (173 amino acids) and basic αB (175 amino acids).[85,86] These two subunits have a sequence homology of approximately 57 per cent. The αA-subunit is lens specific and αB-crystallin, with a molecular weight of 20 kDa, has been detected in a number of nonlenticular tissues, including cardiac tissue and glial cells in the PNS and CNS.[87-89]

Many of the small HSPs are not constitutively expressed and only produced after stress conditions.[80] They selectively bind to nonnative proteins and exhibit chaperone activity, are stress inducible and confer thermoprotection.[90] One special feature of these molecular chaperones is that they effectively trap aggregation-prone folding intermediates, keep them in a refoldable conformation and prevent further aggregation, even at permissive temperature. The stress denatured proteins are then renatured in cooperation with other HSPs in an ATP-dependent manner.[14,15] ATP enhanced the molecular chaperone function of human αB-crystallin[91] and a recent study demonstrates that binding of ATP to the α-crystallin and not its hydrolysis conferred enhanced aggregation prevention ability.[92] At normal physiological concentrations ATP causes the sHSPs to change their conformation and release the substrate proteins, allowing renaturation by HSP70. Also, it was demonstrat..., that in the absence of ATP sHSPs are more effective than HSP70 in protecting other proteins from heat induced denaturation.[15]

Small HSPs contribute to the balance between cell death and survival and can confer resistance against apoptotic stimuli, they associate with the cytoskeleton, specifically with microfilaments and intermediate filaments, have physiological roles as modulators of the cytoskeleton and might protect the cytoskeleton during stress.[5,6,8,21] In the CNS they are predominantly localized to glial cells. Data from our laboratory show that in cell cultures derived from the brains of newborn rats, HSP25 and αB-crystallin are prominently and constitutively expressed in astrocytes, highly inducible in astrocytes and oligodendrocytes, and not detectable in microglia (Stahnke et al, in preparation). In cultured neurons HSP25 and αB-crystallin are inducible expressed, the latter however only weakly (Vollmer et al, in preparation). It might be speculated that the high constitutive levels of sHSPs in astrocytes have protective effects, and are one of the reasons why astrocytes are less sensitive to stress situations than neurons and oligodendrocytes. Increased expression especially of αB-crystallin has been demonstrated in various neurodegenerative disorders, including Alzheimer's disease, Alexander's disease, and Parkinson's disease, and αB-crystallin is a major component of glial filament inclusions.[6,12,80,93] This topic will be discussed in the following chapter by Richter-Landsberg and Goldbaum.

HSP32/ Heme Oxygenase-1, a Sensor of Oxidative Stress

HSP32, also known as heme oxygenase-1 (HO-1), is a small stress protein with enzymatic activity that belongs to the heme oxygenase (HO) family. This family of proteins catalyzes the oxidative degradation of heme to biliverdin. Biliverdin is subsequently converted to bilirubin and equimolar amounts of carbon monoxide and iron. Bilirubin and biliverdin have free radical scavenging capabilities, and thus can act as potent antioxidants. Heme and iron on the other hand can exacerbate intracellular oxidative stress by an increased formation of reactive oxygen intermediates.[94] Three mammalian isoforms of HO have been identified. HO-1, an inducible enzyme that is synthesized in response to heat shock, heme and oxidative stress,[95,96] is most highly concentrated in tissues that are involved in the catabolism of heme proteins.[97] HO-2 represents the constitutively expressed and noninducible isoform, that is present in highest concentrations in the brain and testes and is thought to be particularly involved in signalling pathways.[98] HO-3 is an isoform with low catalytic activity, its physiological roles remain largely obscure.[99] HO-1 has no chaperone activity, but is considered as a sensor and regulator of many forms of oxidative stress, and may be a major player in protecting cells from oxidative stress in association with inflammatory processes and neurodegenerative disorders.[94,95,100] This topic is discussed in Chapter 4 by Calabrese et al in this volume.

In normal brain HO-1/HSP32 expression level is rather low and induced after heat shock or ischemia in neuronal and glial cells.[101-104] After focal cerebral infarctions and traumatic human brain injury the prolonged expression of HO-1 was detected mainly in microglia at the border of the lesion,[105] and HO-1 was transiently induced in astrocytes after cortical stab wound injury.[106] In cell culture studies it has been demonstrated that HO-1 can be induced by hydrogen peroxide in astrocytes and oligodendrocytes.[33,107]

HO-1 is upregulated in a number of neurodegenerative diseases, including PD and AD, which points to a contribution of oxidative stress in the pathogenesis of these neurological disorders.[94] Reactive oxygen and nitrogen species play a role in inflammation and also in inflammatory demyelinating disorders, such as MS.[108] HO-1 expression was significantly increased in experimental autoimmune encephalomyelitis (EAE), a commonly used animal model of MS, and could be localized to infiltrating monocytes[109] and reactive microglia and astrocytes, respectively (Stahnke et al in preparation).[110] Our data show that oligodendrocytes are specifically sensitive to oxidative stress, respond by the onset of programmed cell death and the upregulation of HO-1.[33,111,112] Also, proinflammatory cytokines TNFα and IL-1β, which are upregulated in MS and EAE, caused the upregulation of HO-1 in cultured rat astrocytes followed by iron sequestration by the mitochondrial compartment.[113] Hence, prolonged HO-1 upregulation in glia might be an indicator of the pathological consequences of oxidative stress, exerted by inflammatory processes and/or damage to the blood brain barrier. The sustained upregulation possibly is not cytoprotective and might promote further oxidative stress, which irreversibly leads to cell death and degenerative processes.

Concluding Remarks

HSPs play important roles in the maintenance of physiological processes and are essential during development and differentiation. Induction of HSPs after stressful insults play protective roles and provide tolerance against further stress situations. The stress protein response and HSPs might prevent protein aggregation and support proteolytic degradation by guiding nonreparable proteins to the ubiquitin proteasome system. HSPs help to detoxify proteins and prevent the assembly of protein aggregates with cytotoxic potential. HSPs on the other hand are involved in the regulation of cell death and survival. Severe stress and sustained upregulation might contribute to the onset of programmed cell death or induce necrotic processes. The relative levels of the individual HSPs are important. Cells that express too high or too low levels may exhibit derangement of cell proliferation or developmental abnormalities, respectively. In aging organisms misfolded proteins accumulate leading to the recruitment of free chaperones. Cells might then be depleted of available free chaperones and defects in protein repair, protein transport, signal transduction and cytoskeletal organisation may be caused. Protein aggregates or inclusion bodies, characterized by the presence of HSPs, specific abnormal cytoplasmic proteins and often cytoskeletal components, characteristically occur in neurodegenerative diseases. Stress proteins may serve as biomarkers to identify stress specificity and localize pathological processes leading to cell and organelle damage. HSPs may serve not only as neuropathological markers, but their contribution to disease pathogenesis also makes them a promising target for therapeutic intervention and drug development.

Acknowledgements

This work was supported by grants of the Deutsche Forschungsgemeinschaft, the Gemeinnützige Hertie-Stiftung, and the Society for Progressive Supranuclear Palsy. I thank Dr. Olaf Goldbaum for help with the graphic arts.

References

1. Jaattela M. Heat shock proteins as cellular lifeguards. Ann Med 1999; 31:261-271.
2. Helmbrecht K, Zeise E, Rensing L. Chaperones in cell cycle regulation and mitogenic signal transduction: A review. Cell Prolif 2000; 25:585-621.
3. Rutherford SL, Lindquist S. Hsp90 as a capacitor for morphological evolution. Nature 1998; 396:336-342.
4. Nollen EAA, Morimoto RI. Chaperoning signaling pathways: Molecular chaperones as stress-sensing heat shock proteins. J Cell Sci 2002; 115:2809-2818.
5. Liang P, MacRae TH. Molecular chaperones and the cytoskeleton. J Cell Sci 1997; 110:1431-1440.
6. Head MW, Goldman JE. Small heat shock proteins, the cytoskeleton and inclusion body formation. Neuropath Appl Neurobiol 2000; 26:304-312.
7. Csermely P. A nonconventional role of molecular chaperones: Involvement in the cytoarchitecture. News Physiol Sci 2001; 15:123-126.
8. Richter-Landsberg C, Bauer NG. Tau-inclusion body formation in oligodendroglia: The role of stress proteins and proteasome inhibition. Int J Devl Neurosci 2004; 22:443-451.
9. Söti C, Sreedhart AS, Csermely P. Apoptosis, necrosis and cellular senescence: Chaperone occupancy as a potential switch. Aging cell 2003; 2:39-45.
10. Sreedhar AS, Csermely P. Heat shock proteins in the regulation of apoptosis: New strategies in tumor therapy. Pharmacol Ther 2004; 101:227-257.
11. Fink AL. Chaperone-mediated protein folding. Physiol Rev 1999; 79:425-449.
12. Richter-Landsberg C, Goldbaum O. Stress proteins in neural cells: Functional roles in health and disease. Cell Mol Life Sci 2003; 60:337-349.
13. Lee AS. The glucose-regulated proteins: Stress induction and clinical applications. Trends Biochem 2001; 26:504-510.
14. Muchowski PJ, Hays LG, Yates JR et al. ATP and the core α-crystallin domain of the small heat shock protein αB-crystallin. J Biol Chem 1999; 274:20190-20195.
15. Wang K, Spector A. ATP causes the small heat shock proteins to release denatured protein. Eur J Biochem 2001; 268:6335-6345.
16. Ciechanover A, Brundin P. The ubiquitin proteasome system in neurodegenerative diseases: Sometimes the chicken, sometimes the egg. Neuron 2003; 40:427-445.
17. Forman MS, Trojanowski JQ, Lee VMY. Neurodegenerative diseases: A decade of discoveries paves the way for therapeutic breakthroughs. Nature Medicine 2004; 10:1055-1063.
18. Esser C, Alberti S, Höhfeld J. Cooperation of molecular chaperones with the ubiquitin proteasome system. Biochim Biophys Acta 2004; 1695:171-188.
19. Lee DH, Goldberg AL. Proteasome inhibitors: Valuable new tools for cell biologists. Trends Cell Biol 1998; 8:397-403.
20. Kim D, Kim SH, Li GC. Proteasome inhibitors MG132 and lactacystin hyperphosphorylate HSF1 and induce hsp70 and hsp27 expression. Biochem Biophys Res Comm 1999; 254:264-268.
21. Goldbaum O, Richter-Landsberg C. Proteolytic stress causes heat shock protein induction, tau-ubiquitination and the recruitment of ubiquitin to tau-positive aggregates in oligodendrocytes in culture. J Neurosci 2004; 24:5748-5757.
22. Sherman MY, Goldberg AL. Cellular defenses against unfolded proteins: A cell biologist thinks about neurodegenerative diseases. Neuron 2001; 29:15-32.
23. Pirkkala L, Nykänen P, Sistonen L. Roles of heat shock transcription factors in regulation of the heat shock response and beyond. FASEB J 2001; 15:1118-1131.
24. Morimoto RI. Cells in stress: Transcriptional activation of heat shock genes. Science 1993; 259:1409-1410.
25. Wu C. Heat shock transcription factors: Structure and regulation. Annu Rev Cell Dev Biol 1995; 11:441-469.
26. Zou J, Guo Y, Guettouche T et al. Repression of heat shock transcription factor HSF1 activation by HSP90 (HSP90 complex) that forms a stress-sensitive complex with HSF1. Cell 1998; 94:471-480.
27. Kawazoe Y, Nakai A, Tanabe M et al. Proteasome inhibition leads to the activation of all members of the heat shock factor family. Eur J Biochem 1998; 255:356-362.
28. Kim D, Li GC. Proteasome inhibitors lactacystin and MG132 inhibit the dephosphorylation of HSF1 after heat shock and suppress thermal induction of heat shock genes. Biochem Biophys Res Comm 1999; 264:352-358.
29. Pirkkala L, Alastalo TP, Zuo X et al. Disruption of heat shock factor1 reveals an essential role in the ubiquitin proteolytic pathway. Mol Cell Biol 2000; 20:2670-2675.
30. Dwyer BE, Nishimura RN. Heat shock proteins and neuroprotection in CNS culture. In: Mayer J, Brown I, eds. Heat Shock Proteins in the Nervous System. New York: Academic Press, 1994:101-121.

31. Sharp FR, Massa SM, Swanson RA. Heat shock protein protection. Trends Neurosci 1999; 22:97-99.
32. Yenari MA, Giffard RM, Steinberg GK. The neuroprotective potential of heat shock protein 70 (HSP70). Molecular Med Today 1999; 5:525-531.
33. Goldbaum O, Richter-Landsberg C. Stress proteins in oligodendrocytes: Differential effects of heat shock and oxidative stress. J Neurochem 2001; 78:1233-1242.
34. Picard D. Heat shock protein 90, a chaperone for folding and regulation. Cell Mol Life Sci 2002; 59:1640-1648.
35. Pratt WB, Toft DO. Regulation of signaling by the hsp90/hsp70-based chaperone machinery. Exp Biol Med 2003; 228:111-133.
36. Brown IR. Induction of heat shock genes in mammalian brain by hyperthermia and tissue injury. In: Mayer J, Brown I, eds. Heat Shock Proteins in the Nervous System. New York: Academic Press, 1994:31-53.
37. McClellan AJ, Frydman J. Molecular chaperones and the art of recognizing a lost cause. Nature Cell Biol 2001; 3:E1-E3.
38. Young JC, Moarefi I, Hartl FU. Hsp90: A specialized but essential protein-folding tool. J Cell Biol 2001; 154:267-273.
39. Neckers L. Hsp90 inhibitors as novel cancer chemotherapeutic agents. Trends Mol Med 2002; 8:S55-S61.
40. Sreedhar AS, Kalmar E, Csermely P et al. Hsp90 isoforms: Functions, expression and clinical importance. FEBS Lett 2004; 562:11-15.
41. Grammatikakis N, Vultur A, Ramana CV et al. The role of Hsp90N, a new member of the Hsp90 family, in signal transduction and neoplastic transformation. J Biol Chem 2002; 277:8312-8320.
42. Sreedhar AS, Söti C, Csermely P. Inhibition of Hsp90: A new strategy for inhibiting protein kinases. Biochim Biophys Acta 2004; 1697:233-242.
43. Sidera K, Samiotaki M, Panayotou G et al. Involvement of cell surface HSP90 in cell migration reveals a novel role in the developing nervous system. J Biol Chem 2004; 279:45379-45388.
44. Gerges NZ, Tran IC, Backos DS et al. Independent functions of hsp90 in neurotransmitter release and in the continuous synaptic cycling of AMPA receptors. J Neurosci 2004; 24:4758-4766.
45. Galigniana MD, Harrell JM, Housley PR et al. Retrograde transport of the glucocorticoid receptor in neurites requires dynamic assembly of complexes with the protein chaperone hsp90 and is linked to the CHIP component of the machinery for proteasomal degradation. Mol Brain Res 2004; 123:27-36.
46. Whitesell L, Mimnaugh EG, de Costa B et al. Inhibition of heat shock protein HSP90-pp60v-src heteroprotein complex formation by benzoquinone ansamycins: Essential role for stress proteins in oncogenic transformation. Proc Natl Acad Sci USA 1994; 91:8324-8328.
47. Le Brazidec JY, Kamal A, Busch D et al. Synthesis and biological evaluation of a new class of geldanamycin derivatives as potent inhibitors of Hsp90. J Med Chem 2004; 47:3865-3873.
48. Kamal A, Sensintaffar J, Boehm MF et al. A high-affinity conformation of Hsp90 confers tumour selectivity on Hsp90 inhibitors. Nature 2002; 425:407-410.
49. Neckers L, Lee YS. The rules of attraction. Nature 2003; 425:357-359.
50. Sittler A, Lurz R, Lueder G et al. Geldanamycin activates a heat shock response and inhibits huntingtin aggregation in a cell culture model of Huntington's disease. Human Mol Genet 2001; 10:1307-1315.
51. McLean PJ, Klucken J, Shin Y et al. Geldanamycin induced Hsp70 and prevents α-synuclein aggregation and toxicity in vitro. Biochem Biophys Res Comm 2004; 321:665-669.
52. Lu A, Ran R, Parmentier-Batteur S et al. Geldanamycin induces heat shock proteins in brain and protects against cerebral ischemia. J Neurochem 2002; 81:355-364.
53. Minami Y, Höhfeld J, Ohtsuka K et al. Regulation of heat-shock protein 70 reaction cycle by the mammalian DnaJ homolog, Hsp40. J Biol Chem 1996; 271:19617-19624.
54. Michels AA, Kanon B, Konings AWT et al. Hsp70 and Hsp40 chaperone activities in the cytoplasm and the nucleus of mammalian cells. J Biol Chem 1997; 272:33283-33289.
55. Frydman J, Nimmesgern E, Ohtsuka K et al. Folding of nascent polypeptide chains in a high molecular mass assembly with molecular chaperones. Nature 1994; 370:111-117.
56. Ohtsuka K, Suzuki T. Roles of molecular chaperones in the nervous system. Brain Res Bull 2000; 53:141-146.
57. Nishimura RN, Dwyer BE, Clegg KB et al. Comparison of the heat shock response in cultured cortical neurons and astrocytes. Mol Brain Res 1991; 9:39-45.
58. Nishimura RN, Dwyer BE, Vinters HV et al. Heat shock in cultured neurons and astrocytes: Correlation of ultrastructure and heat shock protein synthesis. Neuropathol Appl Neurobiol 1991; 17:139-147.
59. McCabe T, Simon RP. Hyperthermia induces 72kDa heat shock protein expression in rat brain nonneuronal cells. Neurosci Lett 1993; 159:163-165.

60. Satoh J, Kim SU. HSP72 induction by heat stress in human neurons and glia cells in culture. Brain Res 1994; 653:243-250.
61. D'Souza DD, Antel JP, Freedman MS. Cytokine induction of heat shock expression in human oligodendrocytes: An interleukin-1-mediated mechanism. J Neuroimmunol 1994; 50:17-24.
62. Satoh J, Yamamura T, Kunishita T et al. Heterogeneous induction of 72-kDa heat shock protein (HSP72) in cultured mouse oligodendrocytes and astrocytes. Brain Res 1992; 573:37-43.
63. Satoh J, Kim SU. Constitutive and inducible expression of heat shock protein HSP72 in oligodendrocytes in culture. Neuroreport 1995; 6:1081-1084.
64. Dwyer BE, Nishimura RN, de Vellis J et al. Regulation of heat shock protein synthesis in rat astrocytes. J Neurosi Res 1991; 28:352-358.
65. Foster JA, Brown IR. Differential induction of heat shock mRNA in oligodendrocytes, microglia and astrocytes following hyperthermia. Mol Brain Res 1997; 45:207-218.
66. Almazan G, Liu HN, Khorchid A et al. Exposure of developing oligodendrocytes to cadmium causes HSP72 induction, free radical generation, reduction in glutathione levels, and cell death. Free Radic Biol Med 2000; 29:858-869.
67. Juurlink BHJ. Type-2 astrocytes have much greater susceptibility to heat stress than type-1 astrocytes. J Neurosi Res 1994; 38:196-201.
68. Beere HM, Green DR. Stress management-heat shock protein-70 and the regulation of apoptosis. Trends Cell Biol 2001; 11:6-10.
69. Itoh H, Kobayashi R, Waui H. Mammalian 60 kDa stress protein (chaperonin homolog). Identification, biochemical properties, and localization. J Biol Chem 1995; 270:13429-13435.
70. Brosnan CF, Battistini L, Gao YL et al. Heat shock proteins and multiple sclerosis: A review. J Neuropath Exp Neurol 1996; 55:389-402.
71. Soltys BJ, Gupta RS. Cell surface localization of the 60 kDa heat shock chaperonin protein (hsp60) in mammalian cells. Cell Biol Int 1997; 21:315-320.
72. Maguire M, Coates ARM, Henderson B. Chaperonin 60 unfolds its secrets of cellular communication. Cell Stress Chaperones 2002; 7:317-329.
73. D'Souza SM, Brown IR. Constitutive expression of heat shock proteins Hsp90, Hsc70, Hsp70 and Hsp60 in neural and nonneural tissues of the rat during postnatal development. Cell Stress Chaperones 1998; 3:188-199.
74. Dwyer DS, Liu Y, Miao S et al. Neuronal differentiation in PC12 cells is accompanied by diminished inducibility of Hsp70 and Hsp60 in response to heat and ethanol. Neurochem Res 1996; 21:659-666.
75. Bajramovic JJ, Bsibsi M, Geutskens SB et al. Differential expression of stress proteins in human adult astrocytes in response to cytokines. J Neuroimmunol 2000; 106:14-22.
76. Selmaj K, Brosnan CF, Raine CS. Expression of heat shock protein-65 by oligodendrocytes in vivo and in vitro: Implications for multiple sclerosis. Neurology 1992; 42:795-800.
77. Freedman MS, Buu NN, Ruijs TC et al. Differential expression of heat shock proteins by human glial cells. J Neuroimmunol 1992; 41:231-238.
78. Samali A, Cia JY, Zhivotovsky B et al. Presence of a preapoptotic complex of pro-caspase-3, Hsp60 and Hsp10 in the mitochondrial fraction of jurkat cells. EMBO J 1999; 18:2040-2048.
79. Derham BK, Harding JJ. α-Crystallin as a molecular chaperone. Progr Retinal Eye Res 1999; 18:463-509.
80. Haslbeck M. sHsps and their role in the chaperone network. Cell Mol Life Sci 2002; 59:1649-1657.
81. Jakob U, Buchner J. Assisting spontaneity: The role of Hsp90 and small Hsps as molecular chaperones. Trends Biochem Sci 1994; 19:205-211.
82. Gusev NB, Bogatcheva NV, Marston SB. Structure and properties of small heat shock proteins (sHsp) and their interaction with cytoskeleton proteins. Biochem 2002; 67:511-519.
83. Ciocca DR, Oesterreich S, Chamness GC et al. Biological and clinical implications of heat shock protein 27,000 (HSP27): A review. J Natl Cancer Inst 1993; 85:1558-1570.
84. Ehrnsperger M, Lilie H, Gaestel M et al. The dynamics of Hsp25 quaternary structure. Structure and function of different oligomeric species. J Biol Chem 1999; 274:14867-14874.
85. Horwitz J. Alpha-crystallin can function as a molecular chaperone. Proc Natl Acad Sci USA 1992; 89:10449-10453.
86. DeJong WW, Leunissen JAM, Voorter CEM. Evolution of the alpha-crystallin/small heat-shock protein family. Mol Biol Evol 1993; 10:103-126.
87. Iwaki T, Iwaki A, Goldman JE. Cellular distribution of alpha B-crystallin in nonlenticular tissues. J Histochem Cytochem 1990; 38:31-39.
88. Kato K, Shinohara H, Kurobe N et al. Tissue distribution and developmental profiles of immunoreactive αB-crystallin in the rat determined with a sensitive immunoassay system. Biochimica Biophysica Acta 1991; 1074:201-208.

89. Lowe J, Mcdermott H, Pike I et al. αB-crystallin expression in nonlenticular tissues and selective presence in ubiquitinated inclusion bodies in human disease. J Pathol 1992; 166:61-68.
90. Muchowski PJ, Bassuk JA, Lubsen NH et al. Human alphaB-crystallin. Small heat shock protein and molecular chaperone. J Biol Chem 1997; 272:2578-2582.
91. Muchowski PJ, Clark JI. ATP-enhanced molecular chaperone functions of the small heat shock protein human αB-crystallin. Proc Natl Acad Sci USA 1998; 95:1004-1009.
92. Biswas A, Das KP. Role of ATP on the interaction of α-crystallin with its substrates and its implications for the molecular chaperone function. J Biol Chem 2004; 279:42648-42657.
93. Dabir D, Trojanowski JQ, Richter-Landsberg C et al. Expression of small heat shock protein αB-crystallin in tauopathies with glial pathology. Am J Pathol 2004; 164:143-153.
94. Schipper HM. Heme oxygenase-1: Role in brain aging and neurodegeneration. Exp Gerontology 2000; 35:821-830.
95. Elbirt KK, Bonkovsky HL. Heme oxygenase: Recent advances in understanding its regulation and role. Proc Assoc Amer Phys 1999; 111:438-447.
96. Sharp FR, Massa SM, Swanson RA. Heat shock protein protection. Trends Neurosci 1999; 22:97-99.
97. Shibahara S, Muller R, Taguchi H et al. Cloning and expression of cDNA for rat heme oxygenase. Proc Natl Acad Sci USA 1985; 82:7865-7869.
98. Maines MD, Trakshel GM, Kutty RK. Characterization of two constitutive forms of rat liver microsomal heme oxygenase. Only one molecular species of the enzyme is inducible. J Biol Chem 1986; 261:411-419.
99. McCoubrey Jr WK, Huang TJ, Maines MD. Isolation and characterization of a cDNA from the rat brain that encodes hemoprotein heme oxygenase-3. Eur J Biochem 1997; 247:725-732.
100. Dennery PA. Regulation and role of heme oxygenase in oxidative injury. Curr Topics Cellular Regulation 2000; 36:181-199.
101. Ewing JF, Haber SN, Maines MD. Normal and heat-induced patterns of expression of heme oxygenase-1 (HSP32) in rat brain: Hyperthermia causes rapid induction of mRNA and protein. J Neurochem 1992; 58:1140-1149.
102. Takeda A, Onodera H, Sugimoto A et al. Increased expression of heme oxygenase mRNA in rat brain following transient forebrain ischemia. Brain Res 1994; 666:120-124.
103. Takeda A, Kimpara T, Onodera H et al. Regional difference in induction of heme oxygenase-1 protein following rat transient forebrain ischemia. Neurosci Lett 1996; 205:169-172.
104. Koistinaho J, Miettinen S, Keinanen R et al. Long-term induction of haem oxygenase-1 (HSP-32) in astrocytes and microglia following transient focal brain ischaemia in the rat. Eur J Neurosci 1996; 8:2265-2272.
105. Beschorner R, Adjodah D, Schwab JM et al. Long-term expression of heme oxygenase-1 (HO-1, HSP-32) following focal cerebral infarctions and traumatic brain injury in humans. Acta Neuropathol 2000; 100:377-384.
106. Dwyer BE, Nishimura RN, Lu SY et al. Transient induction of heme oxygenase after cortical stab wound injury. Mol Brain Res 1996; 38:251-259.
107. Dwyer BE, Nishimura RN, Lu SY. Differential expression of heme oxygenase-1 in cultured cortical neurons and astrocytes determined by the aid of a new heme oxygenase antibody. Response to oxidative stress. Mol Brain Res 1995; 30:37-47.
108. Smith KJ, Kapoor R, Felts PA. Demyelination: The role of oxygen and nitrogen species. Brain Pathol 1999; 9:69-92.
109. Schluesener HJ, Seid K. Heme oxygenase-1 in lesions of rat experimental autoimmune encephalomyelitis and neuritis. J Neuroimmunol 2000; 110:114-120.
110. Emerson MR, LeVine SM. Heme oxygenase-1 and NADPH cytochrome P450 reductase expression in experimental allergic encephalomyelitis: An expanded view of the stress response. J Neurochem 2000; 75:2555-2562.
111. Richter-Landsberg C, Vollgraf U. Mode of cell injury and death after hydrogen peroxide exposure in cultured oligodendroglia cells. Exp Cell Res 1998; 244:218-229.
112. Vollgraf U, Wegner M, Richter-Landsberg C. Activation of AP-1 and NF-kappaB transcription factors is involved in hydrogen peroxide-induced apoptotic cell death of oligodendrocytes. J Neurochem 1999; 73:2501-2509.
113. Mehindate K, Sahlas DJ, Frankel D et al. Proinflammatory cytokines promote glial heme oxygenase-1 expression and mitochondrial iron deposition: Implications for multiple sclerosis. J Neurochem 2001; 77:1386-1395.

CHAPTER 2

Small Heat Shock Proteins and the Cytoskeleton:

Their Role in Inclusion Body Formation in Glial Cells

Christiane Richter-Landsberg* and Olaf Goldbaum

Abstract

The integrity of the cytoskeleton is an essential determinant for the function and survival of nerve cells and glia, and hence provides a sensitive target for stress situations. The small heat shock proteins (sHSPs) αB-crystallin and HSP25/27 specifically interact with cytoskeletal elements, i.e., intermediate filaments, microfilaments and microtubules. sHSPs are closely related and act as molecular chaperones, preventing the aggregation of proteins after stress. They are more prominent in glial cells than in neurons, and represent prominent constituents of inclusion bodies originating in astrocytes and oligodendrocytes in neurodegenerative diseases associated with cytoskeletal abnormalities. Hence, sHSPs are particularly important in neurological disorders with glial pathology. In this chapter we focus on the sHSPs, their interaction with the cytoskeleton in neural cells during normal and stressful conditions, and their role in inclusion body formation where they might affect amyloid fibril formation and toxicity.

Introduction

Filamentous protein inclusions, which are observed in nerve cells and glia in a number of neurodegenerative disorders,[1-2] often contain a variety of heat shock proteins (HSPs) in addition to cytoskeletal proteins,[3] indicating that stress situations which impair cytoskeletal integrity and functions contribute to pathogenesis. Furthermore, the inclusions often stain with antibodies against ubiquitin (for review see Chapter 7 by Wyttenbach and Arrigo and Chapter 6 by Pierre et al in this volume), suggesting that the inclusions are targeted for degradation by the proteasomal system. Hence, despite the upregulation of HSPs aggregates are formed, indicating that HSP induction was an unsuccessful attempt to prevent protein denaturation and aggregation and to rescue the cells. Furthermore, the aggregates could not be removed, most likely due to a failure or overload of the protein degradation machinery.

In this chapter we focus on the sHSPs, namely αB-crystallin and HSP27, their interaction with cytoskeletal proteins in neural cells during normal and stressful conditions, and role in inclusion body formation. αB-Crystallin and HSP25/27 are closely related, act as molecular chaperones and prevent the aggregation of other proteins after physiological stress,

*Corresponding Author: Christiane Richter-Landsberg—Department of Biology, Molecular Neurobiology, University of Oldenburg, POB 2503, D-26111 Oldenburg, Germany. Email: Christiane.Richter.Landsberg@Uni-Oldenburg.de

Heat Shock Proteins in Neural Cells, edited by Christiane Richter-Landsberg.

but have only limiting capacities for refolding stressed proteins. The sHSPs tend to form large aggregates and phosphorylation affects their oligomeric state[4] (see also Chapter 1 by Richter-Landsberg, this volume). They are increased in a variety of neurodegenerative disorders, including Alzheimer's disease (AD), Parkinson's disease (PD), multiple system atrophy (MSA), Alexander's disease, and also may play a role in the pathogenesis of multiple sclerosis, as detailed by Brosnan et al (Chapter 8, this volume). sHSPs are incorporated in pathological inclusions and evidence has accumulated that they might affect amyloid fibril formation and toxicity. These sHSPs are generally more abundant in glial cells than in neurons.

The Cytoskeleton in Nerve Cells and Glia

The integrity, dynamic properties and spatial organization of the cytoskeletal network are essential determinants for the function and survival of nerve cells and glia. The cytoskeleton, consisting of microtubules (MTs), intermediate filaments (IFs) and microfilaments (MFs), also called actin filaments, establish and maintain cell morphology and provide the framework for organelle trafficking and cellular sorting processes. Nerve cells, oligodendrocytes and astrocytes can be distinguished by their IF system. Astrocytes are characterized by the presence of glia filaments, which are composed of the intermediate filament protein GFAP (glial fibrillary acidic protein), while mature oligodendrocytes are devoid of IFs.[5-6] In adult human neurons five major IF proteins are observable, namely peripherin (57 kDa), which is mainly found in the peripheral nervous system, α-internexin (66 kDa), which is widely expressed in the CNS, and the three neurofilament proteins (NF-L, 68 kDa; NF-M, 145 kDa; and NF-H, 200 kDa) which copolymerize and form the 10 nm neurofilaments which are most abundant.[7]

Actin is the major constituent of the cytoskeleton of almost all eukaryotic cells, and MFs (actin filaments) are organized into cytoarchitectural meshworks which are the basis for cell motility, cell migration and developmental processes. MFs are involved in neurite growth and guidance[8] and also during the formation of oligodendrocyte processes and branches.[9]

MTs are hollow cylinders composed of heterodimers of α- and β-tubulin, they are organized in orderly arrays in neurons and oligodendrocytes with the fast growing plus-end directed to the distal ends and the slow growing minus-end oriented towards the cell body.[5,10] This polarity is essential for the functions of the microtubule associated motor protein family, e.g., kinesins and dyneins, which move to the plus- or minus-end, respectively, and are involved in the assembly and orientation of cell structures, such as the Golgi apparatus and the endoplasmic reticulum.[11] MTs have special dynamic properties, the dynamic instability is a necessary requirement for their many functions, including cell motility and cell division, organelle transport and cell morphogenesis.[12] MTs originate from variable structures, called microtubule organizing centers (MTOC) or centrosomes lying in discrete regions near the nucleus of cells. At the MTOC a third tubulin isotype, namely γ-tubulin is present, which is thought to be involved in the tubulin nucleation process. Furthermore, MT dynamics and polarity are regulated by a variety of microtubule associated proteins (MAPs). These play a critical role in promoting tubulin assembly and MT stabilization. MAPs copurify with tubulin during repeated cycles of assembly and disassembly, and constitute a diverse family of proteins including high molecular weight MAPs (270-350 kDa) such as MAP1A, MAP1B (MAP5), and MAP2, and low molecular weight MAPs (55-82 kDa), including tau proteins.[13-14] Tau proteins are most abundantly expressed in nerve cells, but also have been identified in oligodendrocytes and their many processes, where they might play similar roles in MT stabilization and contribute to pathological situations.[6,15]

Neurons and oligodendrocytes are characterized by their elaborate MT network and have a high content of tubulin, while in fibrous astrocytes the GFAP IF system is most prominent. Neurons are distinguished into axonal and dendritic compartments, and establish a complex network of long and branched processes which needs to be maintained.[10] Oligodendrocytes produce vast amounts of myelin by spirally enwrapping neuronal axons

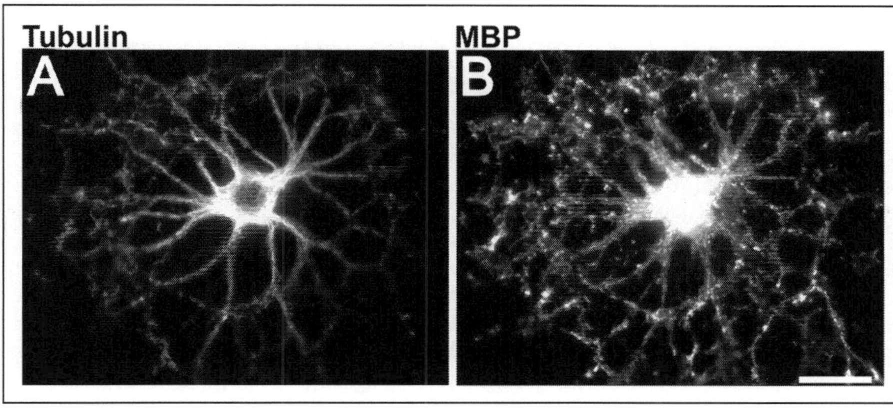

Figure 1. Microtubules in oligodendrocytes. Rat brain oligodendrocytes (7 days in culture) were fixed with 3% paraformaldehyde and indirect immunofluorescence double staining was carried out with anti-tubulin antibodies (A) and anti-myelin basic protein (MBP) antibodies (B). Bar represents 25 µm. Note the elaborate microtubule network in the membranous extensions which are also rich in MBP.

with their membranous processes. During myelin compaction, the cytoplasm is extruded and membranes become closely opposed. Only small cytoplasmic channels remain which are particularly rich in MTs, they provide the template for active transport of cytoplasmic constituents, intracellular sorting processes, positioning of organelles, and are essential for the maintenance of the metabolically active myelin membrane.[5] Oligodendrocytes in culture (Fig. 1) are characterized by their numerous cytoplasmic extensions and flat membranous sheets, these contain an extensive network of MTs.[6]

The individual cytoskeletal elements, although varying in molecular composition, influence each others activities and functions. The cytoskeleton is a major target of stress, heat shock or other stressors can cause the collapse of the filamentous network or induce a loss of stress fibers. The proper synthesis and assembly of cytoskeletal constituents and their maintenance are essential to preserve cell viability and vital for the cells. Hence, protection of the cytoskeleton during stress is a major survival response. A number of molecular chaperones, in particular the sHSPs, interact with cytoskeletal elements in normal cells, influence their function, might assist their proper assembly and spatial organization. They are recruited to cytoskeletal elements and protect the cytoskeleton in response to cellular stress.[4,16-17] In nonstressed cells, molecular chaperones might be a constituent of the cytoarchitecture.[18] In fact, several members of the HSP70 family were initially characterized as microtubule binding proteins, attaching to polymerized tubulin at a site which is also recognized by MAPs.[19]

Interaction of Small Heat Shock Proteins with Cytoskeletal Proteins

The sHSPs interact with all three cytoskeletal elements in cells of the nervous system and are possibly involved in the regulation of muscle and nonmuscle motility.

Microfilaments

Only few studies have been carried out on the role of sHSPs in microfilament formation and protection in neural cells. In adult sensory neurons, a role for HSP27 in neuritogenesis and subsequent neurite extension, possibly be regulating actin cytoskeletal dynamics, has been suggested[20] (see also Chapter 3 by Herbert et al, this volume). HSP25 directly can interact with actin, it predominantly acts as a chaperone and preferentially interacts with thermally unfolded actin, preventing the formation of insoluble aggregates.[21] Furthermore, sHSPs in a phosphorylation-dependent manner have modulatory functions on microfilament polymerization.[4] Murine HSP25 in its unphosphorylated monomeric state inhibits actin assembly, an activity which is not exerted by phosphorylated monomers or multimeric nonphosphorylated complexes of HSP25.[22] Overexpression of HSP27 in CHO fibroblasts attenuates cytochalasin D-induced disruption of microfilaments, an effect which is phosphorylation dependent. Only the nonphosphorylatable form of HSP27 can accelerate the reappearance of microfilaments after treatment with the drug,[23] while in another study it was shown that phosphorylatable HSP27 is protective against oxidative stress, prevents actin fragmentation and dell death.[24] Hence, phosphorylation regulates the effects of HSP25/27 on actin dynamics. Also αB-crystallin has an affinity for actin. In rat glioma cells, using sense/antisense strategy, the presence of additional αB-crystallin confers thermoresistance and a reduction is accompanied by a disorganized microfilament network.[25] Studies on the role of αB-crystallin in rat cardiac myoblast cell line H9c2 further indicate that after proteasomal inhibition by MG-132 or lactacystin, sHSPs are redistributed within the cells and colocalize with the actin cytoskeleton.[26] This association seems to be independent of phosphorylation and is not observable in rat C6 glioma cells. Additionally, in this system proteasomal inhibition leads to an increase of sHSPs in the detergent insoluble fraction and to the appearance of sHSP-positive protein lumps near the microfilaments, possibly indicating that actin filaments are involved in clearing the cytoplasm from harmful aggregates,[26] thereby resembling the role of microtubules in the formation of aggresomes (see below, and the Chapter 6 by Pierre et al, this volume).

Intermediate Filaments

The sHSPs αB-crystallin and HSP25/27 are involved in the maintenance of the IF system, an affinity for desmin, GFAP, vimentin and peripherin has been reported.[27-28] As mentioned before, the sHSPs seem to be more prominent in glia cells and mainly involved in glial pathology.[27,29-30] Upregulation of αB-crystallin stabilizes gliafilaments (GFAP-IF) and after physiological stress, such as heat shock, its binding specificity to IFs is enhanced.[31-32] Overexpression of αB-crystallin by adenoviral gene transfer causes the disaggregation of GFAP cytoplasmic inclusions in astrocytes, which are formed in the cells after transfection of GFAP cDNA.[33] When αB-crystallin is overexpressed in nonstressed primary astrocytes, thick gliafilaments are debundled and a filigree network of finer and more numerous filaments is observable. No change in GFAP solubility was detected.[34] Hence, an increase in αB-crystallin expression, in the absence of stress, can modify intermediate filament organization, a property which is not shared by HSP27.[34]

Microtubules and Microtubule Associated Proteins (MAPs)

While several members of the HSP70 family were characterized as MAPs and their role in cytoskeleton formation was implied (for review see ref. 16), little is known about the potential roles of sHSPs in MT polymerization and maintenance. The association of HSP27 with tubulin and MTs in HeLa cells and of αB-crystallin in L6 myoblast cell lysates was reported.[35-36] Furthermore, tubulin aggregation was prevented through formation of binary complexes with αB-crystallin.[36] In a later study the same authors describe that αB-crystallin increases the resistance of MTs to depolymerization in cells and in vitro.[37] In C6 glioma cells the synthesis and accumulation of αB-crystallin in response to agents that promote the disassembly of MTs, e.g., colchicine, vinblastine and nocodazole, is induced, while taxol, a MT destabilizing drug, causes its suppression.[38] Proteasomal inhibitors, such as MG-132, induce the upregulation of various HSPs, including αB-crystallin and HSP25 in astrocytes, neurons and oligodendrocytes (Richter-Landsberg et al, unpublished data). Cultured astrocytes have a high level of constitutively expressed HSP25 which after proteasomal stress accumulates in small aggregates near the center of the cells (Fig. 2). Immunoblot analysis of

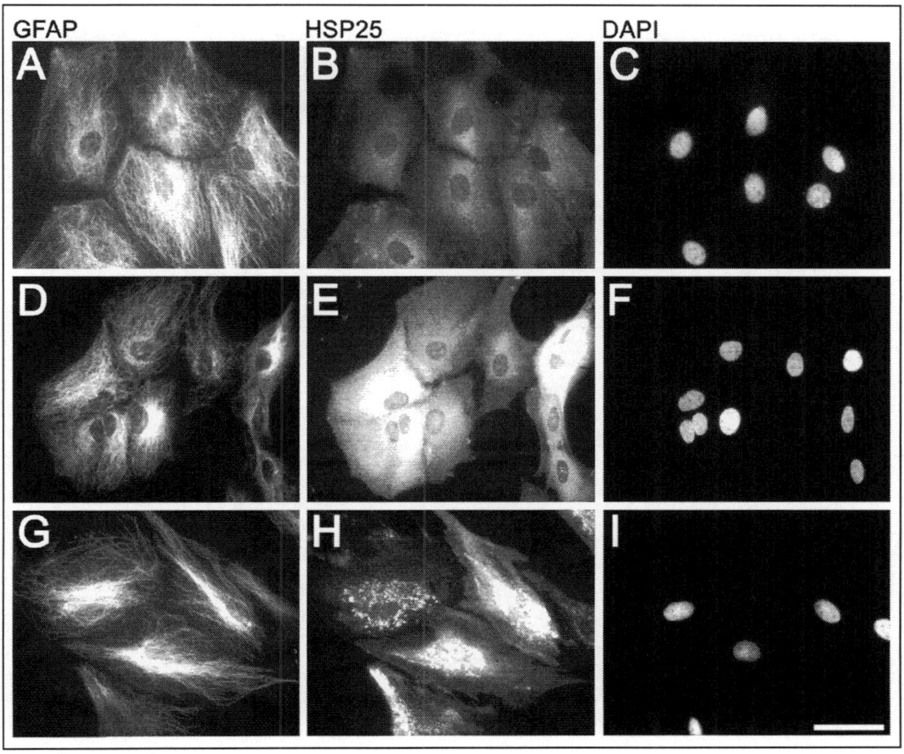

Figure 2. Upregulation and distribution of HSP25 in astrocytes after heat shock and proteasomal stress. Cultured astrocytes, derived from the brains of newborn rats, were fixed with 3% paraformaldehyde and subjected to indirect immunofluorescence staining, using antibodies against intermediate filament protein GFAP (A,D,G) and HSP25 (B,E,H). Cell nuclei were visualized by DAPI staining (C,F,I). Astrocytes were subjected to heat shock (44°C for 30 min) and after 24 h recovery period stained by indirect immunofluorescence (D-F) or treated with the proteasomal inhibitor MG-132 (5 μM) for 24 h (G-I). Untreated astrocytes are seen in A-C. Heat shock and proteasomal inhibition cause the upregulation of HSP25 in astrocytes. After heat shock HSP25 is diffusely distributed in the cytoplasm, while after treatment with MG-132, HSP25-immunoreactivity is mainly concentrated in the cell center and small and larger aggregates are visible. Bar represents 40 μm.

MTs, isolated from astrocytes after stabilization with taxol, indicates that a proportion of the stress-induced sHSPs binds to MTs. Data from our laboratory further demonstrate that also in primary oligodendrocytes derived from rat brain and in the oligodendroglia cell line OLN-93,[39] HSP25 and αB-crystallin bind to MTs after proteasomal inhibition[40-41] (**Fig. 3**). Specifically αB-crystallin interacts with MTs upon stress, is highly inducible and almost completely recruited to the MTs, indicating that it assists in maintaining MT integrity during cellular stress.

HSPs also influence the properties of MAPs, specifically of the low molecular weight MAPs, the tau proteins. Tau is a MAP that binds to the walls of MTs by means of binding domains located in the carboxy terminal region of the protein. As mentioned above, it is important for MT stability and assembly. In the normal human brain six isoforms are expressed, this heterogeneity is further increased by phosphorylation at multiples sites, and

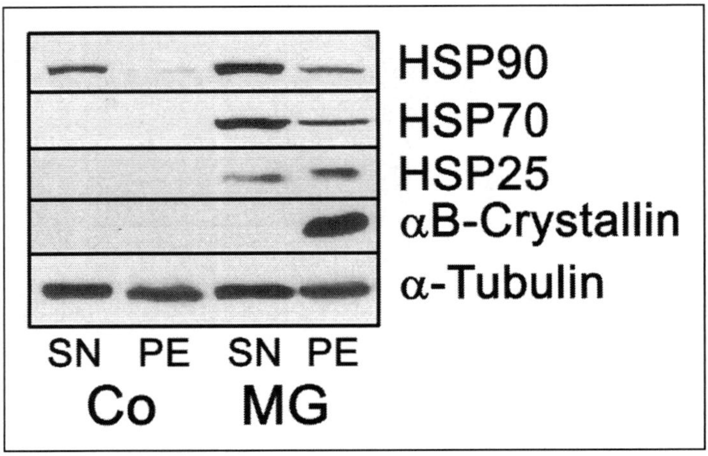

Figure 3. Proteasomal stress leads to the association of αB-Crystallin and HSP25 with microtubules in oligodendrocytes. Oligodendrocytes (7 days in culture) were either untreated (Co) or treated with the proteasomal inhibitor MG-132 (1 μM, 24 h). Thereafter, microtubules (MT) were isolated after stabilization with Taxol as described (Goldbaum and Richter-Landsberg).[40] The presence of HSPs was determined in the pellet (PE, microtubule fraction) and supernatant (SN, cytoplasmic soluble fraction) by Western blot procedure. It is seen that after proteasomal stress (MG) αB-crystallin is only detectable in the MT fraction (PE), and that HSP25 is associated with MTs but also present in the supernatant (SN). HSP90 and HSP70 in contrast thereto are mainly present in the supernatant.

hyperphosphorylation negatively affects MT-binding properties.[14] The chaperones HSP90 and HSP70 increase the affinity of tau protein with microtubules, as was shown in several cell culture model systems including primary neurons.[42] Increased levels of HSP90 and HSP70, exerted by incubation with the antibiotic geldanamycin, promote the solubility of tau, cause a reduction in the basal level of tau phosphorylation, and a direct association of the chaperones with tau. Furthermore, two other studies suggest that CHIP (carboxyl terminus of the HSC70 interacting protein), a cochaperone of HSP70 possessing intrinsic E3 ubiquitin ligase activity, interacts with tau and is able to rescue cells from phosphorylated tau-induced toxicity.[43-44] CHIP/HSP70 was suggested to regulate tau ubiquitination, degradation and aggregation.[44] In oligodendrocytes, proteasomal inhibition by MG-132 causes the association of tau with αB-crystallin and tau ubiquitination is promoted.[40] In human brain tissue, HSP27 preferentially binds pathological hyperphosphorylated tau isolated from AD brains, but not unphosphorylated tau isolated from normal brains. Binding of HSP27 to hyperphosphorylated tau alters the conformation and reduces its concentration by facilitating its degradation and phosphorylation.[45] Hence, cell-type and stress-specific upregulation of sHSPs not only might play an important role in regulating the dynamic properties and stability of cytoskeletal elements, but also is involved in the stabilization or elimination of associated proteins which are linked to neurodegenerative disorders and are components of filamentous inclusions found in a variety of diseases.

sHSPs and Inclusion Body Formation in Glial Cells

Misfolded proteins assemble and accumulate in aggregates before the onset of cell degeneration, and it is likely that this is a mechanism initially aimed at removing toxic material to rescue cells. At later stages of the disease, the growing aggregates may confer death signals leading to neural degeneration, as the number and regional distribution of the inclusions correlate with the clinical picture. Intracellular filamentous inclusions are the characteristic hallmark of AD and PD, similar inclusions are found in diverse other neurodegenerative diseases, such as Pick's disease (PiD), corticobasal degeneration (CBD), progressive supranuclear palsy (PSP), frontotemporal dementia with parkinsonism linked to chromosome 17 (FTDP-17), dementia with Lewy bodies (DLB) and multiple system atrophy (MSA).[2,46] In tauopathies, such as AD, CBD, PSP, FTDP-17, tau is the major constituent of the inclusions, while in synucleinopathies, such as PD, DLB and MSA, inclusions are characterized by the presence of α-synuclein (see also Chapter 7 by Wyttenbach and Arrigo, this volume). The function of α-synuclein is not well defined.[2,46] Like tau, it is a natively unfolded protein with little ordered secondary structure. α-Synuclein is concentrated in nerve terminals, but also present in cultured oligodendrocytes where it is downregulated during differentiation.[5] Evidence is emerging indicating a considerable overlap of the pathological and clinical features of patients with tauopathies and synucleinopathies, pointing to a mechanistical link between these disorders.[47] This is substantiated by the fact that tau and α-synuclein can associate and promote fibrillogenesis of each other in vitro and in transgenic mouse models.[47-48] Recently it has been reported that α-synuclein comprises chaperone-like activity,[49] shares physical and functional homology with 14-3-3 proteins, a family of chaperone proteins particularly abundant in the brain,[50] and also shares sequence homologies with sHSPs.[51] As such, α-synuclein might protect cells from thermal and oxidative stress and suppress aggregation of substrate proteins in an ATP-dependent manner.[49,51-52] Hence, depending on the cellular context, α-synuclein may protect cellular proteins from denaturation, and when overexpressed might contribute to the accumulation of protein aggregates and to cell death in neurodegenerative disease. In this respect it is interesting to note that αB-crystallin was shown to be a potent in vitro inhibitor of α-synuclein fibrillization and redirects α-synuclein from fibril formation towards an amorphous aggregation pathway.[53]

Tau-positive and α-synuclein-positive inclusions are present in nerve cells and glia, pointing to the fact that not only neurons, but also astrocytes and oligodendrocytes play a role in the pathogenesis of neurodegenerative diseases.[54] While in AD and PD intracellular protein inclusions are preferentially formed in neurons, in FTDP-17 tau-pathology is present in neurons and glia, and tau-positive inclusions originating in oligodendrocytes are specifically prominent in PSP, CBD, and PiD.[55] Morphologically diverse tau-positive astrocytic lesions also occur, specifically tufted astrocytes in PSP and astrocytic plaques in CBD serve as diagnostic tool.[56] Oligodendroglia lesions and white matter pathology are characterized by the occurrence of glial fibrillary tangles (GFT), coiled bodies and glial cytoplasmic inclusions (GCIs). GFTs and coiled bodies are mainly observed in PSP, CBD, and PiD.[55,57] GFTs occurring in oligodendrocytes contain hyperphosphorylated tau, but in contrast to neurofibrillary tangles (NFTs) in neurons, they do not form solid filaments and do not associate with ubiquitin.[58] GCIs are round or crescent shaped inclusions in the cytoplasm of oligodendroglia, they are the histological hallmark of MSA, a specific adult onset degenerative disease of the nervous system, characterized by varying degrees of Parkinsonism, cerebellar ataxia, and autonomic dysfunction.[59-61] Signature neuronal inclusions in PD are termed Lewy bodies, which are comprised of fibrils assembled from α-synuclein. Similarly, α-synuclein is the major building block of GCIs, which are also immunoreactive for ubiquitin.[59,61]

GCIs in MSA stain intensely with antibodies against ubiquitin and αB-crystallin and are positive for α- and β-tubulin, while variable reports indicate that they are tau-positive.[57,60,62] In Rosenthal fibres, the characteristic inclusions in astrocytes of patients with Alexander's disease, αB-crystallin, HSP27 and ubiquitin coassemble with GFAP.[27] HSP27 and αB-crystallin are also induced in Creutzfeldt-Jakob disease and AD.[63-65] The specific association of sHSPs with senile plaques in AD but not with NFTs was observed,[66] while in an earlier study αB-crystallin positive neurons were described. Their density correlated with the density of NFTs and it was suggested that αB-crystallin might be involved in the late, severe stage of pathogenesis.[67] Our data further show that αB-crystallin but not HSP70 is mainly expressed in tauopathies with glial pathology.[68] This enhanced expression of αB-crystallin seems to be specific for disorders with prominent glial pathology, suggesting distinct mechanisms for tau aggregation in neurons versus glia in neurodegenerative disorders.

The molecular mechanisms leading to inclusion body formation and to cell death in oligodendrocytes are far from understood. Several studies show that protein inclusions are formed, when the proteasome activity is decreased. Small aggregates can be delivered to the nascent inclusion body by an active, retrograde transport system along the microtubules, which have been termed aggresomes[69] (see also Chapter 6 by Pierre et al, this volume). This process requires an intact cytoskeleton and is inhibited by MT destabilizing drugs.[41,70-71] Aggresomes are dynamic structures and distinct from inclusion bodies found in neurodegenerative disorders, which most likely represent an end stage of the protein aggregation process and contribute to cell death. In cultured oligodendrocytes, stress proteins can be upregulated by proteasomal inhibition and tau can be induced to accumulate and form inclusions that resemble fibrillary deposits similar to those observed in neurodegenerative diseases.[17,40] Oligodendroglia inclusion bodies in these cell culture models contain αB-crystallin and ubiquitin, and can be stained by Thioflavin-S, a histochemical dye specifically binding to crossed β-pleated sheet structures that is used to identify fibrillary protein aggregates (Fig. 4). Generally, oligodendrocytes are specifically sensitive to stress situations and respond by the onset of programmed cell death after heat shock, oxidative and proteasomal stress.[40,72-75] Astrocytes are much more resistant, and we hypothesize that this is due to their high level of endogenously expressed HSP25 and αB-crystallin, which is further enhanced in response to stress (Richter-Landsberg et al, in preparation).

In summary, individual cell types in the CNS show different susceptibility to stress, and the correct balance of the levels of HSPs seems to be important for cell survival, rescue mechanisms and to attenuate toxic effects of amyloid proteins. Chronic overexpression of sHSPs, instead of protecting the cells, might further contribute to the onset of cell death.

Concluding Remarks

sHSPs display chaperone activity, they undergo dynamic assembly into large aggregates and their oligomerization is required for chaperone function and substrate binding. They can specifically interact with components of the cytoskeleton and under stress conditions might preserve cell shape and integrity. The cytoskeleton is a specific target for stress situations and disturbed in many neurodegenerative disorders. As molecular chaperones, sHSPs protect cells by preventing irreversible aggregation of proteins during stress, and are implicated in several diseases not only as protecting, but also as causative agents. Nerve cells and glia have different thresholds for stress responses, and sHSPs seem to be particularly important in neurological diseases with glial pathology. They are prominent constituents in inclusion bodies originating in astrocytes and oligodendrocytes, which are a classical feature of CNS diseases associated with neuronal cytoskeletal abnormalities, and are also expressed during lesional development in multiple sclerosis. A better understanding of their cell-type and stress-specific regulation and how they are involved in disease progression is a major challenge in the field.

Acknowledgements

The authors acknowledge support of their research by the Deutsche Forschungsgemeinschaft, the Gemeinnützige Hertie-Stiftung, and the Society for Progressive Supranuclear Palsy.

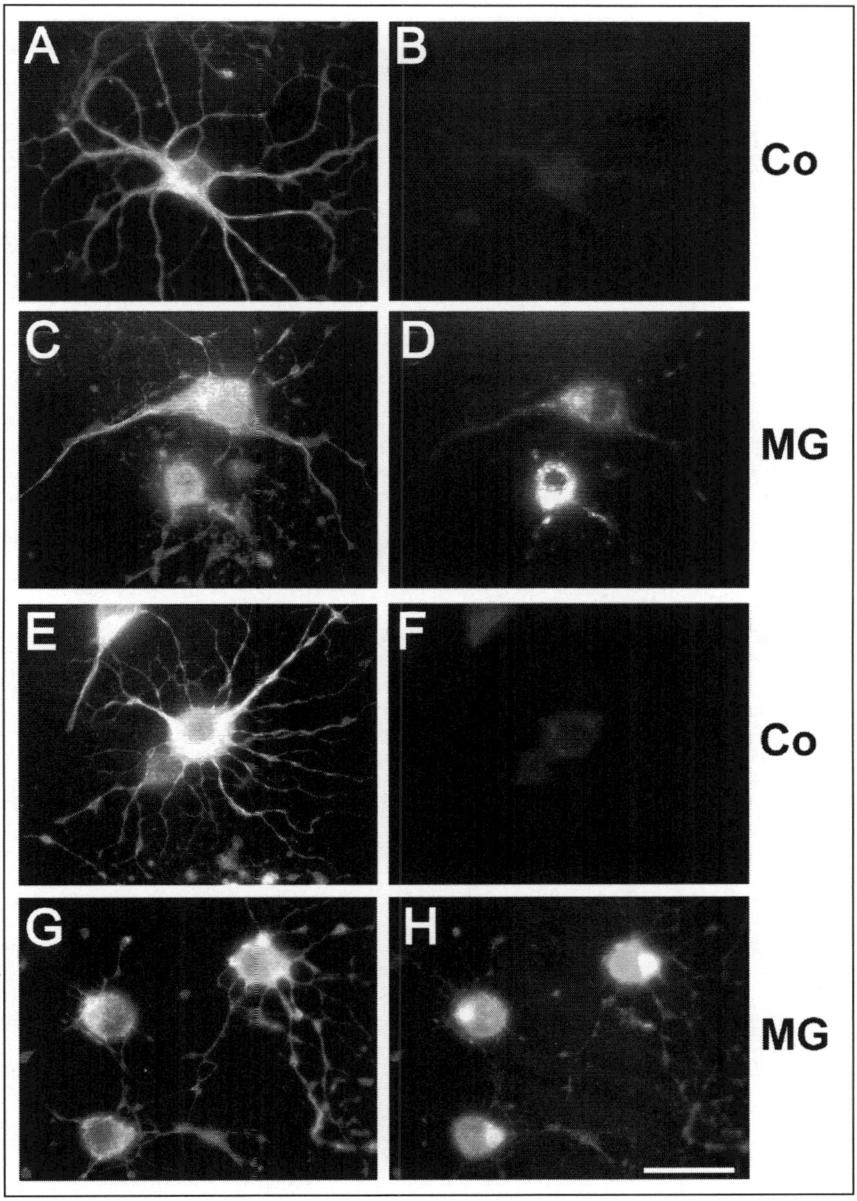

Figure 4. Proteasomal stress leads to the formation of thioflavin-S-positive protein aggregates containing αB-crystallin in oligodendrocytes. Rat brain oligodendrocytes (7 days in culture) either were untreated (Control, A,B and E,F) or treated with MG-132 (1 μM, 24 h) (C,D and G,H). Cells were then fixed with 3% paraformaldehyde, and subjected to indirect immunofluorescence staining using anti-αB-crystallin antibodies (B,D) and anti-tubulin antibodies (A,C) or with anti-tubulin antibodies (E,G) followed by staining with thioflavin-S (F,H). Bar represents 25 μm.

References

1. Goedert M, Spillantini MG, Davies SW. Filamentous nerve cell inclusions in neurodegenerative diseases. Curr Opin Neurobiol 1998; 8:619-632.
2. Forman SF, Trojanowksi JQ, Lee VMY. Neurodegenerative diseases: A decade of discoveries paves the way for therapeutic breakthroughs. Nature Med 2004; 10:1055-1063.
3. Richter-Landsberg C, Bauer NG. Tau-inclusion body formation in oligodendroglia: The role of stress proteins and proteasome inhibition. Int J Devl Neurosci 2004; 22:443-451.
4. Gusev NB, Bogatcheva NV, Marston SB. Structure and properties of small heat shock proteins (sHsp) and their interaction with cytoskeleton proteins. Biochem 2002; 67:511-519.
5. Richter-Landsberg C. The oligodendroglia cytoskeleton in health and disease. J Neurosci Res 2000; 59:11-18.
6. Richter-Landsberg C. Organization and functional roles of the cytoskeleton in oligodendrocytes. Microsc Res Tech 2001; 52:628-636.
7. Cairns NJ, Lee VMY, Trojanowski JQ. The cytoskeleton in neurodegenerative disease. J Pathol 2004; 204:438-449.
8. Meyer G, Feldman EL. Signaling mechanisms that regulate actin-based motility processes in the nervous system. J Neurochem 2002; 83:490-503.
9. Song J, Goetz BD, Baas PW. Cytoskeletal organization during the formation of oligodendrocyte processes and branches. Mol Cell Neurosci 2001; 17:624-636.
10. Brandt R. Cytoskeletal mechanisms of neuronal degeneration. Cell Tissue Res 2001; 305:255-265.
11. Hirokawa N. Kinesin and dynein superfamily proteins and the mechanism of organelle transport. Science 1998; 279:519-526.
12. Valiron O, Caudron N, Job D. Microtubule dynamics. Cell Mol Life Sci 2001; 58:2069-2084.
13. Sanchez C, Diaz-Nido J, Avila J. Phosphorylation of microtubule-associated protein 2 (MAP2) and its relevance for the regulation of the neuronal cytoskeleton function. Progr Neurobiol 2000; 61:133-168.
14. Buée L, Bussière T, Buée-Scherrer V et al. Tau protein isoforms, phosphorylation and role in neurodegenerative disorders. Brain Res Rev 2000; 33:95-130.
15. Gorath M, Stahnke T, Mronga T et al. Developmental changes of tau protein and mRNA in cultured rat brain oligodendrocytes. Glia 2001; 36:89-101.
16. Liang P, MacRae TH. Molecular chaperones and the cytoskeleton. J Cell Sci 1997; 110:1431-1440.
17. Richter-Landsberg C, Goldbaum O. Stress proteins in neural cells: Functional roles in health and disease. Cell Mol Life Sci 2003; 60:337-349.
18. Csermely P. A nonconventional role of molecular chaperones: Involvement in the cytoarchitecture. News Physiol Sci 2001; 16:123-126.
19. Sanchez C, Padilla R, Paciucci R et al. Binding of heat shock protein 70 (hsp70) to tubulin. Arch Biochem Biophys 1994; 310:428-432.
20. Williams KL, Rahimtula M, Mearow KM. Hsp27 and axonal growth in adult sensory neurons in vitro. BMC Neuroscience 2005; 6:24.
21. Panasenko OO, Kim MV, Marston SB et al. Ineraction of the small heat shock prtoein with molecular mass 25kDa (hsp25) with actin. Eur J Biochem 2003; 270:892-901.
22. Benndorf R, Hayeß K, Ryazantsev S et al. Phosphorylation and supramolecular organization of murine small heat shock protein HSP25 abolish its actin polymerization-inhibiting activity. J Biol Chem 1994; 269:20780-20784.
23. Lavoie JN, Lambert H, Hickey E et al. Modulation of cellular thermoresistance and actin filament stability accompanies phosphorylation-induced changes in the oligomeric structure of heat shock protein 27. Mol Cell Biol 1995; 15:505-516.
24. Huot J, Houle F, Spitz DR et al. HSP27 phosphorylation-mediated resistance against actin fragmentation and cell death induced by oxidative stress. Cancer Res 1996; 56:273-279.
25. Iwaki T, Iwaki A, Tateishi J et al. Sense and antisense modification of glial αB-crystallin production results in alterations of stress fiber formation and thermoresistance. J Cell Biol 1994; 125:1358-1393.
26. Verschuure P, Cores Y, van den Ijssel PR et al. Translocation of small heat shock proteins to the actin cytoskeleton upon proteasomal inhibition. J Mol Cell Cardiol 2002; 34:117-128.
27. Head MW, Goldman JE. Small heat shock proteins, the cytoskeleton and inclusion body formation. Neuropath Appl Neurobiol 2000; 26:304-312.
28. Sun Y, MacRae TH. Small heat shock proteins: Molecular structure and chaperone function. Cell Mol Life Sci 2005; 62:2460-2476.
29. Van Rijk AF, Bloemendal H. Alpha-B-crystallin in neuropathology. Ophthalmologica 2000; 214:7-12.

30. Sun Y, MacRae TH. The small heat shock proteins and their function in human disease. FEBS J 2005b; 272:2613-2627.
31. Djabali K, de Nechaud B, Landon F et al. αB-crystallin ineracts with intermdiate filaments in response to stress. J Cell Sci 1997; 110:2759-2769.
32. Muchowski PJ, Valdez MM, Clark JI. αB-crystallin selectively targets intermediate filament proteins during thermal stress. Invest Ophthalmol Vis Sci 1999; 40:951-958.
33. Koyama Y, Goldman JE. Formation of GFAP cytoplasmic inclusions in astrocytes and their disaggregation by αB-crystallin. Am J Pathol 1999; 154:1563-1572.
34. Head MW, Hurwitz L, Kegel K et al. αB-crystallin regulates intermediate filament organization in situ. NeuroReport 2000; 11:361-365.
35. Hino M, Kurogi K, Okubo MA et al. Small heat shock protein 27 (HSP27) associates with tubulin/microtubules in HeLa cells. Biochem Biophys Res Comm 2000; 271:164-169.
36. Arai H, Atomi Y. Chaperone activity of αB-crystallin suppresses tubulin aggregation through complex formation. Cell Struct Funct 1997; 22:539-544.
37. Fujita Y, Ohto E, Katayama E et al. αB-crystallin-coated MAP microtubule resists nocodazole and calcium-induced disassembly. J Cell Sci 2003; 117:1719-1726.
38. Kato K, Ito H, Inaguma Y et al. Synthesis and accumulation of αB-crystallin in C6 glioma cells is induced by agents that promote the disassembly of microtubules. J Biol Chem 1996; 271:26989-26994.
39. Richter-Landsberg C, Heinrich M. OLN-93: A new permanent oligodendroglia cell line derived from primary rat brain glial cultures. J Neurosci Res 1996; 45:161-173.
40. Goldbaum O, Richter-Landsberg C. Proteolytic stress causes heat shock protein induction, tau-ubiquitination and the recruitment of ubiquitin to tau-positive aggregates in oligodendrocytes in culture. J Neurosci 2004; 24:5748-5757.
41. Bauer NG, Richter-Landsberg C. Recruitment of αB-Crystallin and ubiquitin to aggresome-like inclusions in oligodendroglial cells after proteolytic stress: The role of microtubules. J Mol Neurosci 2006, (In press).
42. Dou F, Netzer WJ, Tanemura K et al. Chaperones increase association of tau protein with microtubules. Proc Natl Acad Sci USA 2003; 100:721-726.
43. Shimura H, Schwartz D, Gygi SP et al. CHIP-Hsc-70 complex ubiquitinates phosphorylated tau and enhances cell curvival. J Biol Chem 2004; 279:4869-4876.
44. Petrucelli L, Dickson D, Kehoe K et al. CHIP and Hsp70 regulate tau ubiquitination, degradation and aggregation. Human Mol Gen 2004; 13:703-714.
45. Shimura H, Miura-Shimura Y, Kosik KS. Binding of tau to heat shock protein 27 leads to decreased concentration of hyperphosphorylated tau and enhanced cell survival. J Biol Chem 2004; 279:17957-17962.
46. Goedert M. The significance of tau and α-synuclein inclusions in neurodegenerative diseases. Curr Opinion Gen Dev 2001; 11:342-351.
47. Lee VMY, Giasson BI, Trojanowski JQ. More than just two peas in a pod: Common amyloidogenic properties of tau and α-synuclein in neurodegenerative diseases. Trend Neurosci 2004; 27:129-134.
48. Giasson BI, Forman MS, Higuchi M et al. Initiation and synergistic fibrillization of tau and alpha-synulcein. Science 2003; 300:636-640.
49. Souza JM, Giasson BI, Lee VMY et al. Chaperone-like activity of synucleins. FEBS Lett 2000; 474:116-119.
50. Ostrerova N, Petrucelli L, Farrer M et al. α-Synuclein shares physical and functional homology with 14-3-3 proteins. J Neurosci 1999; 19:5782-5791.
51. Kim TD, Choi E, Rhim H et al. α-Synuclein has structural and functional similarities to small heat shock proteins. Biochem Biophys Res Comm 2004; 324:1352-1359.
52. Kim TD, Paik SR, Yang CH et al. Structural changes in alpha-synuclein affect its chaperone-like activity in vitro. Protein Sci 2000; 9:2489-2496.
53. Rekas A, Adda CG, Aquilina JA et al. Interaction of the molecular chaperone αB-crystallin with α-synuclein: Effects on amyloid fibril formation and chaperone activity. J Mol Biol 2004; 340:1167-1183.
54. Miller DW, Cookson MR, Dickson DW. Glial cell inclusions and the pathogenesis of neurodegenerative diseases. Neuron Glia Biol 2004; 1:13-21.
55. Komori T. Tau-positive glial inclusions in progressive supranuclear palsy, corticobasal degeneration and Pick's disease. Brain Pathol 1999; 9:663-679.
56. Komori T, Arai N, Oda M et al. Astrocytic plaques and tufts of abnormal fibers do not coexist in corticobasal degeneration and progressive supranuclear palsy. Acta Neuropathol 1998; 96:401-408.
57. Chin SSM, Goldman JE. Glial inclusions in CNS degenerative diseases. J Neuropathol Exp Neurol 1996; 55:499-508.

58. Ikeda K, Akiyama H, Arai T et al. Glial tau pathology in neurodegenerative diseases: Their nature and comparison with neuronal tangles. Neurobiol Aging 1998; 19:S85-91.
59. Dickson DW, Lin WL, Liu WK et al. Multiple system atrophy: Asporadic synucleinopathy. Brain Pathol 1999; 9:721-732.
60. Lantos PL. The definition of multiple system atrophy: A review of recent developments. J Neuropathol Exp Neurol 1998; 57:1099-1111.
61. Burn DJ, Jaros E. Multiple system atrophy: Cellular and molecular pathology. Mol Pathol 2001; 54:419-426.
62. Murayama S, Arima K, Nakazato Y et al. Immunocytochemical and ultrastrutural studies of neuronal and oligodendroglial cytoplasmic inclusions in multiple system atrophy. 2. Oligodendroglial cytoplasmic inclusions. Acta Neuropathol 1992; 84:32-38.
63. Renkawek K, de Jong WW, Merck KB et al. αB-crystallin is present in reactive glia in Creutzfeld-Jakob disease. Acta Neuropathol (Berlin) 1992; 83:324-327.
64. Renkawek K, Voorter CE, Brosnan GJ et al. Expression of αB-crystallin in Alzheimer's disease. Acta Neuropathol (Berlin) 1994; 87:155-160.
65. Shinohara H, Inaguma Y, Goto S et al. Alpha B crystallin and HSP28 are enhanced in the cerebral cortex of patients with Alzheimer. J Neurol Sci 1993; 119:203-208.
66. Wilhelmus MM, Boelens WC, Otte-Holler I et al. Small heat shock protein HspB8: Its distribution in Alzheimer's disease brains and its inhibition of amyloid-beta protein aggregation and cerebrovascular amyloid-beta toxicity. Acta Neuropathol (Berlin) 2006; 111:139-149.
67. Mao JJ, Katayama S, Watanabe C et al. The relationship between αB-crystallin and neurofibrillary tangles in Alzheimer's disease. Neuropathol Appl Neurobiol 2001; 27:180-188.
68. Dabir D, Trojanowski JQ, Richter-Landsberg C et al. Expression of small heat shock protein αB-crystallin in tauopathies with glial pathology. Am J Pathol 2004; 164:143-153.
69. Kopito RR. Aggresomes, inclusion bodies and protein aggregation. Trends Cell Biol 2000; 10:524-530.
70. Johnston JA, Ward CL, Kopito RR. Aggresomes: A cellular response to misfolded proteins. J Cell Biol 1998; 143:1883-1898.
71. Garcia-Mata R, Gao YS, Sztul E. Hassles with taking out the grabage: Aggravating aggresomes. Traffic 2002; 3:388-396.
72. Vollgraf U, Wegner M, Richter-Landsberg C. Activation of AP-1 and NF-kappaB transcription factors is involved in hydrogen peroxide-induced apoptotic cell death of oligodendrocytes. J Neurochem 1999; 73:2501-2509.
73. Goldbaum O, Richter-Landsberg C. Stress proteins in oligodendrocytes: Differential effects of heat shock and oxidative stress. J Neurochem 2001; 78:1233-1242.
74. Mronga T, Stahnke T, Goldbaum O et al. Mitochondrial pathway is involved in hydrogen-peroxide-induced apoptotic cell death of oligodendrocytes. Glia 2004; 46:446-455.
75. Goldbaum O, Vollmer G, Richter-Landsberg C. Proteasome inhibition by MG-132 induces apoptotic cell death and mitochondrial dysfunction in cultured rat brain oligodendrocytes but not in astrocytes. Glia 2006; 53:891-901.

Chapter 3

The Role of Hsps in Neuronal Differentiation and Development

Kate Reed Herbert, Afshin Samali and Adrienne Gorman*

Abstract

Heat shock proteins (Hsps) are expressed during development of the nervous system in a temporally and spatially controlled pattern that does not appear to be linked to activation of heat shock transcription factors. The distinct patterns of Hsp expression suggest that they perform unique roles during development and neuronal differentiation. Studies indicate that these proteins can inhibit programmed cell death, regulate cytoskeletal dynamics during neurite outgrowth and axon pathfinding, as well as interact with and regulate intracellular signaling molecules that are involved in neuronal differentiation. Overall, through their involvement in these various processes, expression of Hsps favours neuronal differentiation and survival and, as such, Hsps are emerging as important regulators of the delicate balance between cell death and survival/differentiation during development of the nervous system.

Introduction

Formation of the vertebrate nervous system begins very early in the life of an embryo during gastrulation, as the blastula (a hollow ball of several hundred unspecialized cells) becomes a three-layered embryo. The nervous system arises from the most external of these layers, the ectoderm. The first step in development of the nervous system, termed neural induction, is when part of the ectoderm thickens to form the neural plate and the cells in this region differentiate to become neuronal precursors (also called neuroblasts). This is followed by neurulation, whereby the neural plate bends inwards, so that the folds eventually meet in the dorsal midline to pinch off and form the hollow neural tube. The anterior portion of the closed neural tube expands to form three brain vesicles, that eventually become the forebrain (prosencephalon), the midbrain (mesencephalon) and the hindbrain (rhombencephalon), while the remainder of the tube develops into the spinal cord. In this way, the neural tube directly gives rise to the entire central nervous system. The peripheral nervous system originates largely from the neural crest, a ridge of cells that arises during neurulation, at the interface between the neural and epidermal ectoderm, which then migrate downward to form the dorsal root ganglia of the spinal nerves and some other cell types. As the embryonic brain develops, the neuronal precursor cells proliferate in the germinal zone of the ventricles. When they become post-mitotic they migrate to their final destination in the nervous system, the axon elongates and these newly formed neurons make synapses with target cells. During development, excessive numbers of neurons are produced and those that do not form successful synaptic connections with, or receive trophic support from, target cells undergo programmed cell death whereby they die by apoptosis.[1]

*Corresponding Author: Adrienne Gorman—Department of Biochemistry, National Centre for Biomedical Engineering Science, National University of Ireland, Galway, Ireland. Email: adrienne.gorman@nuigalway.ie

Heat Shock Proteins in Neural Cells, edited by Christiane Richter-Landsberg.

During development of the nervous system individual neuronal precursors and immature neurons undergo extensive morphological changes as they differentiate, migrate, and extend neurites including the axon. These changes are associated with alterations in protein expression. In this chapter, we will focus on the expression of heat shock proteins (Hsps) during normal development of the nervous system and particularly during neuronal differentiation. Much is known about Hsp expression in various neuropathologies and under conditions that cause cellular stress. However, less attention has focused on the expression of these proteins during normal brain development and neuronal differentiation.

Hsp Expression during Development of the Nervous System

During the past decade or so, examination of the distribution of various Hsps in the nervous system has revealed that they are expressed during development (Table 1).[2-10] The mouse is the best characterized model of mammalian embryonic development. Therefore, while studies have been done using rats, drosophila and even human tissue, most of the studies examining Hsp expression during development, including that of the nervous system, have been performed in the mouse. Neurulation in the mouse starts at about embryonic day 7.5 (E7.5) to E8. At E8.5-9.5 neural tube closure and formation of the three primary brain vesicles occurs. By E12 the various brain regions are more clearly defined. The vertebrate brain is still developing during the early postnatal period and therefore, we will include this in our overview of current knowledge regarding Hsp expression during development of the nervous system.

Hsp25 (the mouse homologue of rat and human Hsp27) has been shown to exhibit both transient and sustained patterns of expression during brain development.[3,4,6,8] It is detectable in neuronal precursors at E9.5 in the mouse and is also highly expressed in the first differentiating neurons of the peripheral layer.[3] As development progresses Hsp25 expression is seen throughout the brain.[3] Expression of Hsp25 during development generally colocalizes with the neuronal markers Tuj1 or βIII-tubulin and in cells that express these markers, Hsp25 is very abundant in the axons and dendrites.[3] However, at E12.5 Hsp25 immunostaining also colocalizes with a marker for radial glial cells, along which the immature neurons that are generated at the ventricular surface of the neural tube begin to migrate.[3] Others have also observed occasional staining of astrocytes with Hsp25 antibodies at postnatal day 3 (P3) in the developing mouse brain[4] and in the adult rat brain.[4,11] Later in embryonic development, Hsp25 expression is limited to small groups of neuronal cells and disappears by E17.5.[3]

Expression of Hsp60 has not been determined during embryonic development of the nervous system. However, the protein is expressed in tissue derived from the brainstem, cerebellum and cerebral hemispheres at P1 and the levels increase markedly (by about 25-fold) by P20 and into adulthood.[2] Immunocytochemical studies demonstrate a neuronal localisation of Hsp60 at all stages of postnatal development examined (P1, P15 and adult).[2]

Hsc70 (the constitutive member of the 70 kDa family of Hsps) is normally present at high levels during neural induction and neural tube closure.[3] It is expressed early during neural crest cell migration and follows the differentiation of the first neural tracts.[3] During development of the nervous system, it is found to colocalize with βIII tubulin, indicating its preferential localization to neurons.[2,3,12] Hsc70 continues to be expressed during later stages of brain development. Western blot analysis and immunohistochemistry indicate that Hsc70 is also expressed in the cerebellum and cerebral hemispheres at P1 and the levels are maintained through to the adult brain.[2] In fact, high levels of Hsc70 are detected in neurons in the adult mammalian central nervous system[7,9,13-15] and it accounts for 2-3% of total cellular protein of the adult rat spinal cord.[15]

In contrast to other Hsps, the expression of Hsp70 (the inducible member of the 70 kDa family of Hsps) is relatively restricted in the developing mouse embryo.[3] It is not detectable until E15.5, and at both E15.5 and E17.5 the specific pattern of Hsp70 staining reflects distribution of glial fibrillary acidic protein (GFAP), indicating that it is mainly localized in glial cells.[3] Western blot analysis of the cerebellum and cerebral hemispheres indicate low levels of Hsp70 in

Table 1. Expression of Hsps during development of the nervous system

Hsp	Species	Expression Pattern	Activities	Possible Role in Neuronal Development
All Hsps			Molecular chaperone	Accommodation of increased demand for protein folding during cell growth
Hsp25/Hsp27	Mouse	Embryonic: differentiating neurons from E9.5 (neural tube closure)[3,6,8]	Inhibition of apoptosis	Selection of neuronal precursor cells to die or differentiate[48]
	Rat	Postnatal[4]	Actin binding	Regulation of microfilament assembly/disassembly during neurite outgrowth
	Drosophlia	Larval[10]	Akt binding	Promotion of activation and prolongation of active state to promote neuronal survival and neurite outgrowth
	Human	Second trimester[9]	Tubulin binding	Regulation of microtubule polymerization and neurite outgrowth
Hsp60	Rat	Neuronal, increases during postnatal development[2]	Mitochondrial protein	Mitochondrial biogenesis linked to high metabolic demands of neurons
Hsc70	Mouse	Embryonic: neuronal, high during neuronal induction and neural tube closure[3,6]	Inhibition of apoptosis	Selection of neuronal precursor cells to die or differentiate[47]
	Rat	Postnatal [2]	ATP-dependent clathrin-uncoating enzyme	Fast axonal protein transport
	Human (Hsp73)	Embryonic[7,9]	Tubulin binding	Regulation of microtubule polymerization and neurite outgrowth
Hsp70	Mouse	Embryonic; glial, detectable at late stages (E15.5)[3]	Inhibition of apoptosis	Selection of neuronal precursor cells to die or differentiate
	Rat	Postnatal[2]	Akt binding	Unknown
Hsp90 (α/β)	Mouse	Embryonic[3,6]	Tubulin binding	Regulation of microtubule polymerization and neurite outgrowth
	Rat	Postnatal[2]	Akt binding	Promotion of activation and prolongation of active state to promote neuronal survival and neurite outgrowth

tissues taken from rats at P1 to P15, with a small increase at P20.[2] Cellular examination of Hsp70 was not performed in postnatal tissue due to the low levels detected.[2] However, given the mainly glial localisation of this protein in the embryonic mouse[3] and the adult rat,[16] it seems likely that Hsp70 expression is also restricted to glia during early postnatal development.

At early stages of embryonic development (E12.5), Hsp90β (the constitutive member of the 90 kDa Hsps) is ubiquitously expressed in the CNS, while the inducible form, Hsp90α, is not detected.[3] Hsp90α begins to be expressed at E15.5, with a distribution that is similar to that of Hsp90β, though at much lower levels.[3] Hsp90 expression is fairly constant during postnatal development, although it decreases slightly at P20 in the cerebellum and the cerebral hemispheres.[2] Immunocytochemical studies demonstrate a neuronal localisation of Hsp90 at all stages of postnatal development examined (P1, P15 and adult).[2] In the adult nervous system, Hsp90s account for 1-2% of total cellular protein.[17]

Transcriptional Regulation of Hsp Expression during Neuronal Development and Differentiation

In mammals, induction of Hsps occurs via the activation of three specific transcription factors, HSF1, HSF2 and HSF4, which are members of the winged helix-turn-helix family of transcription factors.[18] Of these, HSF1 is normally expressed in the cytoplasm or the nucleus as an inert monomer, and its activation is mainly associated with induction of Hsps in response to stress.[19] In contrast, HSF2 regulates Hsp expression under nonstress conditions, and has particularly been associated with development and differentiation.[20] The function of HSF4 is unclear, but it may be involved in negative regulation of Hsp gene expression and also, like HSF1, may play a role during stress responses.[21]

HSF1 expression in the developing postnatal rat brain increases from P1 to P30 in the cerebellum and cerebral cortex during which time HSF2 levels decrease.[22] Throughout development, HSF1 expression is mainly nuclear in both neuronal and glial cells, while HSF2 expression is nuclear only during the earlier developmental stages (P2) and then changes to predominantly cytoplasmic, including expression in both the cell body and the dendritic processes by P30.[22] Significantly, however, this profile shows no obvious correlation with Hsp levels during postnatal development.[22] Temporal and tissue-specific HSF2 activation occurs during mouse embryogenesis, with highest expression observed between E8 and E15 and after this time HSF2 activity is found only in the developing central nervous system.[23] Again, however, the localization of HSF2 does not parallel the sites of Hsp70 expression in the nervous system.[23] Expression of both HSF1 and HSF2 occurs during neural induction.[24] HSF2 knockout mice develop normally and the adults display normal cognitive and psychomotor function, indicating that HSF2 is dispensible for normal brain development.[25] HSF1 knockout mice, on the other hand, exhibit prenatal lethality along with numerous other developmental defects.[26] This is intriguing, since HSF1 activation is normally associated with cellular stress. Clearly, the exact role and contribution of HSFs in controlling Hsp expression during neuronal development remains to be elucidated.

It is conceivable that transcription factors other than HSFs and/or other promoter elements may also regulate Hsp expression during development. An example is Brn3a, a POU transcription factor, which has recently been shown to regulate Hsp27 expression.[27] Brn3a is known to promote retinal ganglion cell differentiation,[28] and to control soma size and axon pathfinding of sensory neurons.[29,30] Taken together, these observations suggest that this transcription factor might play a role in the developmental regulation of Hsp27 in these neurons.[31]

Possible Roles of Hsps in Neuronal Differentiation and Development of the Nervous System

Individual Hsps display distinct, but often overlapping, patterns of expression in the developing brain.[2,3] It is likely therefore, that individual Hsps perform unique roles at different stages of neuronal maturation. Little is known about what these roles are, but some probable

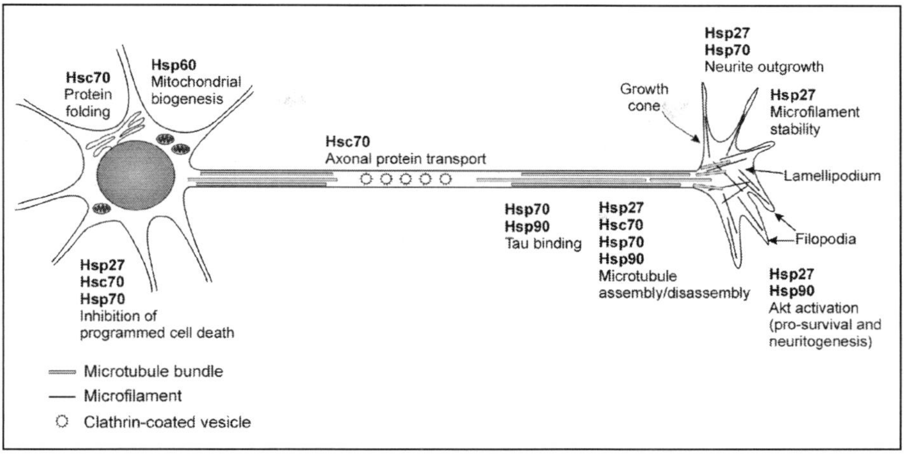

Figure 1. Processes in neuronal development that are affected by Hsps.

functions can be inferred from their known biological activities in nonneuronal cells (see Table 1 for a summary).

The main role of Hsps is to act as molecular chaperones and, in particular, constitutively expressed Hsps are active during protein synthesis to help in the correct folding of proteins and to transport proteins across intracellular membranes. It is therefore probable that Hsp expression during neuronal development and differentiation is at least in part due to the high level of protein synthesis that occurs during this process. Although this idea goes some way to explaining their increased expression during development, it does not explain the different patterns of expression. Control of protein-protein interactions underlies many of the other specific functions of Hsps that may also be involved in neuronal development and differentiation. These include control of cytoskeletal dynamics, inhibition of apoptosis, and regulation of intracellular signaling molecules (see Fig. 1). In accordance with the varied temporal and spatial expression of Hsps observed during neuronal development, it seems likely that Hsps contribute to the developmental process in numerous ways, utilizing their varied functions.

Hsps as Inhibitors of Apoptosis during Development of the Nervous System

One suggested function of inducible Hsps during embryonic development is to provide protection against embryonic damage by heat and other stresses at vulnerable stages of development.[24] Immature or developing neurons are known to display greater sensitivity to induction of cell death by apoptosis than mature neurons.[32-34] It is well known that certain Hsps, particularly Hsp70 and Hsp27, can inhibit apoptosis,[35] including neuronal apoptosis (see Chapter 5 by FitzGerald et al, this volume,). Mechanisms of protection range from direct interference in the apoptotic program, to acting as chaperones for unfolded or misfolded proteins, increasing the cell's antioxidant defences and maintenance of the cytoskeleton.[36,37] This topic is covered in more detail by FitzGerald et al in Chapter 5 in this volume.

Although developing neurons are sensitive to damage-induced death, programmed cell death of about 50% of neurons is a normal, and indeed essential, feature of development of the nervous system.[1,38-40] Neuronal death occurs in 50% of projecting neurons because of their dependence on target-derived trophic factors, combined with the limited supply of these factors as postulated by the neurotrophic factor hypothesis.[41] Whether Hsps can prevent this form of death is not certain, although one study shows that Hsp27 overexpression reduces death of PC12 cells caused by withdrawal of the neurotrophin, nerve growth factor (NGF).[42] There is also evidence that at earlier stages of development, death of 50-70% of neuronal precursor cells occurs during neurogenesis (i.e., when the population is rapidly proliferating).[43,44] Hsc70 may be one factor controlling the balance of cell death and survival during this period.[44-46] Downregulation of Hsc70 expression using antisense leads to an increase in apoptosis in neural tissue from the neurulating chick embryo, demonstrating that this protein is important in promoting survival of neuronal precursors.[47] Hsp27 may also regulate the selection of certain neuronal precursors to differentiate rather than die. In in vitro cultures of rat olfactory neural progenitor cells treated with dopamine, some cells undergo apoptosis while others, which accumulate Hsp27, survive and differentiate.[48] A reduction in Hsp27 expression by antisense reduces the frequency of differentiation.[48] These authors suggest that those cells that differentiate escape from apoptosis by transiently accumulating high levels of Hsp27. It is tempting to speculate that these in vitro findings could be related to the in vivo temporal and spatial pattern of Hsp expression observed in the developing nervous system, and that in the developing nervous system one function of Hsps could be to regulate the selection of certain neuronal precursor cells to die or to differentiate.

Control of the Cytoskeleton by Hsps during Neuronal Differentiation and Neurite Outgrowth

Developing neurons undergo extensive architectural remodeling during their proliferation, migration and differentiation. This remodeling necessitates substantial reorganization of the cytoskeleton, which is composed of microtubules, microfilaments, and intermediate filaments. Together these proteins give the cell its overall shape, and their dynamic reorganization is a key factor in cell motility, organelle transport and cell division.[49] One of the most remarkable features of developing neurons is the outgrowth of neurites from the cell body to form elaborate dendritic arborization and the often highly elongated axon. Clearly, this process involves the formation of a supportive cytoarchitecture with precise spatial organization to facilitate the morphological development of the cell into a mature, functional neuron.

The direction and pattern of neurite growth is determined by the growth cone, a highly motile structure at the end of growing axons and dendrites. The basic mechanism of growth of axons and dendrites is essentially the same, although research has mainly focused on axonal growth. At the leading edge of the growth cone, finger-like protrusions called filopodia extend forward to explore the environment ahead with trailing sheets of membrane (called lamellopodia) in between. The movement of the growth cone is largely based on the presence of dynamic microfilaments and microtubules. Microfilaments and microtubules are composed of globular protein subunits, actin and tubulin, respectively, which enables the rapid assembly/disassembly and reorganization of these cytoskeletal components. The filopodia contain parallel bundles of microfilaments, while the lamellopodia contain networks of crossed filaments that radiate back from the filopodial tips. Stable microtubules are prominent proteins in axons and dendrites and the extension of these microtubules is what lengthens the neurite.[50] Mature axons also contain intermediate filaments which are added relatively late in development. These comprise fibrous proteins that provide mechanical strength and help stabilize cell shape. Five major types of intermediate filament proteins are expressed in adult neurons: three neurofilaments, peripherin and α-internexin,[51] while glial cells express GFAP and vimentin.[52] Neurofilaments are responsible for the radial growth of an axon and thus determine axonal diameter, which is related to the speed at which it conducts impulses.

By direct binding to cytoskeletal proteins, Hsps could significantly influence the cytoskeleton during neuronal differentiation.[49,53,54] Hsp27 was first identified as an actin-binding protein[55] and it can cap the ends of actin filaments, preventing their oligomerisation. Upon Hsp27 phosphorylation, actin is released, allowing for its polymerisation and reorganization.[54-57] Hsp27 binding to actin has been implicated in the differentiation of nonneuronal cells.[54,58] This function of Hsp27 might also be employed during neuritogenesis. In fact, Hsp27 is upregulated in cultured rat adult dorsal root ganglion (DRG) neurons 24 h after they are put in culture and Hsp27 staining is observed throughout the neuron, including the growth cone.[59] A more recent study shows that Hsp27 colocalises with actin and tubulin in filopodia, lamellopodia and focal contacts during early stages of neurite outgrowth of DRG neurons.[60] At later stages, Hsp27 colocalises with actin in processes, branch points, and growth cones. A normal growth pattern is dependent on Hsp27 phosphorylation, which may be attributable to the ability of Hsp27 to regulate microfilament stability. Further evidence to indicate that Hsp27 is a key molecule in neurite outgrowth and axon elongation comes from the observation that, after axotomy in rat DRG neurons, Hsp27 is upregulated and is distributed along axons in vivo and in the growth cone in vitro, which is compatible with a possible role in the growing tip of the axon.[59] Taken together with the selective expression of Hsp27 in migrating and differentiating neurons,[3] these findings suggest that Hsp27 may play a role in neuritogenesis and neurite extension through its ability to regulate microfilaments and thus, alter cell morphology.

Transient expression of Hsp27 is observed during differentiation of immortalized rat olfactory neuronal progenitor 13.S.1.24 cells and a reduction in the levels of Hsp27 expression by antisense was found to reduce the frequency of differentiation.[48] Although these authors suggest that Hsp27 expression prevents apoptosis to allow differentiation to occur, their data do not rule out a role for Hsp27 in the differentiation process per se. Increased expression of Hsp27 is also observed during nerve growth factor (NGF)-induced differentiation of PC12 cells (Fig. 2). In both the rat olfactory neurons[48] and the PC12 cells, no increase in the expression of Hsp70 was detected, indicating that the process of differentiation does not, of itself, induce a classical stress response. Increased Hsp27 expression during neuronal differentiation is not a consistent finding, however. An increase in Hsp25 in mouse P19 embryonic carcinoma cells is observed when they were differentiated to a cardiac but not a neuronal phenotype.[61]

A number of Hsps, including Hsp27,[62] Hsc70,[63] Hsp70[64] and Hsp90[65] can directly bind to tubulin and thereby regulate microtubule polymerisation and movement.[49] Hsp70 can also regulate tubulin polymerization by acting as an antagonist of microtubule-associated proteins (MAPs), thus inhibiting the tubulin polymerization.[64] Also, Hsp70 and Hsp90 can bind the MAP tau, promoting its association with microtubules.[66,67] MAPs may help regulate microtubule assembly, cross-link microtubules to other cytoskeletal components and even help to anchor microtubules to membranes. The levels of the various MAPs change in characteristically different ways during neuronal development, and thus, Hsps could differentially affect MAP binding to microtubules depending on the developmental stage.

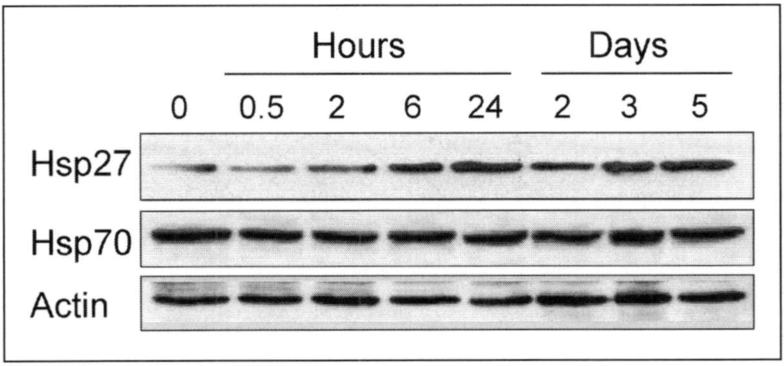

Figure 2. Induction of Hsp27 by NGF treatment of PC12 cells. PC12 cells were treated for 0-5 days with NGF (100 ng/ml). Samples containing 25 μg protein were subjected to 12% SDS-PAGE and then transferred to nitrocellulose membrane which was probed with rabbit poloyclonal antibody against Hsp25 or mouse monoclonal against Hsp70 (both from StressGen Biotechnologies) or rabbit polyclonal against Actin (Sigma).

Members of the Hsp70, Hsp90 and small Hsp families can interact with numerous intermediate filaments.[49] For example, Hsp27 is associated with both GFAP and vimentin in astrocytoma cells.[68] These glial intermediate filaments play a role in glial cell extension along neurites.[69] In neurons, Hsp27 has also been shown to be involved in the organisation of the neurofilament network.[68] In addition, Hsc70 associates with intermediate filaments including vimentin[70,71] although, unlike Hsp27, it does not appear to have any effect on filament dynamics.[68]

It will be interesting to await further developments in this area, as Hsps, through their interaction with cytoskeletal proteins, are likely to play key roles at all stages of neuronal development including neurogenesis, migration of immature neurons, neurite outgrowth and projection of axons towards target cells.

Hsp Interaction with Intracellular Signaling Pathways Involved in Neuronal Differentiation

During development neurotrophins support the differentiation and survival of neurons. The best characterized of these, NGF, mediates its biological effects through binding to TrkA and p75 receptors on the cell membrane. Upon NGF binding to TrkA, autophosphorylation of the receptor activates phosphatidylinositol 3-kinase (PI3K)/Akt and MAPK signaling pathways.[72] Significantly, Hsps have been associated with both of these signaling pathways.[42,73-77] Hsp27 has been found to bind to Akt, a downstream effector of PI3K signaling.[42,73-75,78] Coimmunoprecipitation of Hsp25 with Akt in cytosolic fractions derived from spinal motor neurons following nerve injury, indicates that these two molecules also form complexes in neuronal cells.[74] Hsp interaction with Akt has been reported to promote Akt activation [42,73,75,78] and to prolong its phosphorylated (i.e., active) state.[42] Hsp90 also modulates the activity of Akt by preventing its dephosphorylation and sustaining the active state.[77] Since neurotrophin-induced PI3K/Akt signaling is mainly associated with promoting cell survival during neuronal differentiation[79-83] and is also involved in neuritogenesis in PC12 cells,[84] particularly in suppressing neurite branching,[85] the effect of Hsp27 and Hsp90 may be to positively influence survival and differentiation of neurons during development. In addition, Hsp70 binds to unphosphorylated Akt[86] although the significance of this is unknown.

NGF signaling through TrkA also leads to activation of ERK and p38 MAPK pathways.[87] One of the downstream targets of p38 MAPK is MAPKAP kinase 2,[88-90] which phosphorylates Hsp27.[76] As mentioned previously, Hsp27 phosphorylation is involved in actin reorganization during differentiation, providing a link from neurotrophin to Hsp-controlled cytoskeletal remodeling, via MAPK signaling.

In addition, activation of HSF1 following heat shock in SH-SY5Y neuroblastoma cells is dependent on PI3K activity.[91] This is related to the ability of PI3K activation to inhibit glycogen synthase kinase 3β (GSK3β), a negative regulator of HSF1.[92] PI3K signaling therefore offers a potential signaling mechanism whereby neurotrophins could lead to activation of HSF1, via inhibition of GSK3β, during differentiation. Of course, as described earlier, the exact contribution made by HSF1 to Hsp expression during development or differentiation remains unclear.

Axonal Transport

Formation of dendrites and axonal processes during neuronal development leads to and requires an increase in protein transport. Fast axonal transport involves shuttling of proteins within clathrin-coated vesicles.[93] Hsc70 is an essential molecule in this process as it acts as an ATP-dependent uncoating enzyme that releases clathrin from coated vesicles.[94] During neuronal differentiation, Hsc70 levels may therefore increase in order to facilitate the increased demand for fast axonal transport of correctly folded proteins through clathrin-coated vesicles to the plasma membrane.[95,96] Hsp70, on the other hand, may play a role in slow axonal transport, which involves association of proteins with microtubule, microfilament and neurofilament proteins. Hsp70 may also act as a cytoskeletal matrix cross-linker molecule, linking transported cellular molecules to the actin microfilament via binding of Hsp70 directly to the cellular targeting sequence.[97]

Mitochondrial Biogenesis

Hsp60 is a mitochondrial protein and therefore, increases in its expression probably reflect increases in the mitochondrial content of cells. Mitochondrial biogenesis occurs postnatally in the rat brain[98,99] and thus the increase in Hsp60 levels, and its neuronal distribution during development[2] are probably linked to the rapid growth of neurons and their high metabolic demands.

Concluding Remarks

Hsps are developmentally regulated in the nervous system. However, we are far from having a complete understanding of their roles during neuronal development and differentiation. Clearly, these proteins are important in control of the cytoskeleton, which underpins development of the elaborate processes that extend throughout the nervous system, and whose dynamic regulation is a key element in maintaining cell morphology, organelle trafficking and cellular sorting processes in mature neurons. In particular, Hsp27 is strongly linked to neurite outgrowth and process extension. Hsps may also play a role in regulating the decision of cells to undergo apoptosis or to differentiate during development. The correct balance between cell death and cell differentiation is essential for optimal development of the complex network of neurons within the nervous system. In addition, some Hsps interact with key intracellular signaling pathways that play important roles in neuronal survival and differentiation. Although numerous direct connections between Hsps and neuronal development and differentiation have therefore been identified, the exact relevance and the ultimate consequence of such connections remain unclear. Also, questions concerning regulation of Hsps during development are yet to be fully answered. This exciting area of research clearly holds much scope for future work and offers promise to expand our knowledge of these multi-functional proteins.

Acknowledgements

The authors are grateful to Dr. Una Fitzgerald for critical reading and assistance with figures. The authors' work was supported financially by Enterprise Ireland, the Health Research Board of Ireland, the Higher Education Authority of Ireland and the Millennium Research Fund (NUI, Galway).

References

1. Oppenheim RW. Cell death during development of the nervous system. Annu Rev Neurosci 1991; 14:453-501.
2. D'Souza SM, Brown IR. Constitutive expression of heat shock proteins Hsp90, Hsc70, Hsp70 and Hsp60 in neural and nonneural tissues of the rat during postnatal development. Cell Stress Chaperones 1998; 3(3):188-199.
3. Loones MT, Chang Y, Morange M. The distribution of heat shock proteins in the nervous system of the unstressed mouse embryo suggests a role in neuronal and nonneuronal differentiation. Cell Stress Chaperones 2000; 5(4):291-305.
4. Armstrong CL, Krueger-Naug AM, Currie RW et al. Constitutive expression of heat shock protein HSP25 in the central nervous system of the developing and adult mouse. J Comp Neurol 2001; 434(3):262-274.
5. Walsh DA, Li K, Speirs J et al. Regulation of the inducible heat shock 71 genes in early neural development of cultured rat embryos. Teratology 1989; 40(4):321-334.
6. Loones MT, Rallu M, Mezger V et al. HSP gene expression and HSF2 in mouse development. Cell Mol Life Sci 1997; 53(2):179-190.
7. Kato M, Mizuguchi M, Takashima S. Developmental changes of heat shock protein 73 in human brain. Brain Res Dev Brain Res 1995; 86(1-2):180-186.
8. Gernold M, Knauf U, Gaestel M et al. Development and tissue-specific distribution of mouse small heat shock protein hsp25. Dev Genet 1993; 14(2):103-111.

9. Aquino DA, Padin C, Perez JM et al. Analysis of glial fibrillary acidic protein, neurofilament protein, actin and heat shock proteins in human fetal brain during the second trimester. Brain Res Dev Brain Res 1996; 91(1):1-10.
10. Pauli D, Tonka CH, Tissieres A et al. Tissue-specific expression of the heat shock protein HSP27 during Drosophila melanogaster development. J Cell Biol 1990; 111(3):817-828.
11. Plumier JC, Armstrong JN, Landry J et al. Expression of the 27,000 mol. wt heat shock protein following kainic acid-induced status epilepticus in the rat. Neuroscience 1996; 75(3):849-856.
12. Brown IR, Rush SJ. Expression of heat shock genes (hsp70) in the mammalian brain: Distinguishing constitutively expressed and hyperthermia-inducible mRNA species. J Neurosci Res 1990; 25(1):14-19.
13. Manzerra P, Rush SJ, Brown IR. Temporal and spatial distribution of heat shock mRNA and protein (hsp70) in the rabbit cerebellum in response to hyperthermia. J Neurosci Res 1993; 36(4):480-490.
14. Foster JA, Rush SJ, Brown IR. Localization of constitutive and hyperthermia-inducible heat shock mRNAs (hsc70 and hsp70) in the rabbit cerebellum and brainstem by nonradioactive in situ hybridization. J Neurosci Res 1995; 41(5):603-612.
15. Aquino DA, Klipfel AA, Brosnan CF et al. The 70-kDa heat shock cognate protein (HSC70) is a major constituent of the central nervous system and is up-regulated only at the mRNA level in acute experimental autoimmune encephalomyelitis. J Neurochem 1993; 61(4):1340-1348.
16. Pavlik A, Aneja IS, Lexa J et al. Identification of cerebral neurons and glial cell types inducing heat shock protein Hsp70 following heat stress in the rat. Brain Res 2003; 973(2):179-189.
17. Quraishi H, Brown IR. Expression of heat shock protein 90 (hsp90) in neural and nonneural tissues of the control and hyperthermic rabbit. Exp Cell Res 1995; 219(2):358-363.
18. Ahn SG, Liu PC, Klyachko K et al. The loop domain of heat shock transcription factor 1 dictates DNA-binding specificity and responses to heat stress. Genes Dev 2001; 15(16):2134-2145.
19. Morimoto RI. Regulation of the heat shock transcriptional response: Cross talk between a family of heat shock factors, molecular chaperones, and negative regulators. Genes Dev 1998; 12(24):3788-3796.
20. Calabrese V, Scapagnini G, Ravagna A et al. Molecular chaperones and their roles in neural cell differentiation. Dev Neurosci 2002; 24(1):1-13.
21. Pirkkala L, Nykanen P, Sistonen L. Roles of the heat shock transcription factors in regulation of the heat shock response and beyond. FASEB J 2001; 15(7):1118-1131.
22. Brown IR, Rush SJ. Cellular localization of the heat shock transcription factors HSF1 and HSF2 in the rat brain during postnatal development and following hyperthermia. Brain Res 1999; 821(2):333-340.
23. Rallu M, Loones M, Lallemand Y et al. Function and regulation of heat shock factor 2 during mouse embryogenesis. Proc Natl Acad Sci USA 1997; 94(6):2392-2397.
24. Walsh D, Li Z, Wu Y et al. Heat shock and the role of the HSPs during neural plate induction in early mammalian CNS and brain development. Cell Mol Life Sci 1997; 53(2):198-211.
25. McMillan DR, Christians E, Forster M et al. Heat shock transcription factor 2 is not essential for embryonic development, fertility, or adult cognitive and psychomotor function in mice. Mol Cell Biol 2002; 22(22):8005-8014.
26. Xiao X, Zuo X, Davis AA et al. HSF1 is required for extra-embryonic development, postnatal growth and protection during inflammatory responses in mice. EMBO J 1999; 18(21):5943-5952.
27. Farooqui-Kabir SR, Budhram-Mahadeo V, Lewis H et al. Regulation of Hsp27 expression and cell survival by the POU transcription factor Brn3a. Cell Death Differ 2004; 11(11):1242-1244.
28. Liu W, Khare SL, Liang X et al. All Brn3 genes can promote retinal ganglion cell differentiation in the chick. Development 2000; 127(15):3237-3247.
29. Eng SR, Gratwick K, Rhee JM et al. Defects in sensory axon growth precede neuronal death in Brn3a-deficient mice. J Neurosci 2001; 21(2):541-549.
30. Huang EJ, Liu W, Fritzsch B et al. Brn3a is a transcriptional regulator of soma size, target field innervation and axon pathfinding of inner ear sensory neurons. Development 2001; 128(13):2421-2432.
31. Xiang M, Gan L, Zhou L et al. Targeted deletion of the mouse POU domain gene Brn-3a causes selective loss of neurons in the brainstem and trigeminal ganglion, uncoordinated limb movement, and impaired suckling. Proc Natl Acad Sci USA 1996; 93(21):11950-11955.
32. Vyas S, Juin P, Hancock D et al. Differentiation-dependent sensitivity to apoptogenic factors in PC12 cells. J Biol Chem 2004; 279(30):30983-30993.
33. Wright KM, Linhoff MW, Potts PR et al. Decreased apoptosome activity with neuronal differentiation sets the threshold for strict IAP regulation of apoptosis. J Cell Biol 2004; 167(2):303-313.

34. Benn SC, Woolf CJ. Adult neuron survival strategies—slamming on the brakes. Nat Rev Neurosci 2004; 5(9):686-700.
35. Samali A, Cotter TG. Heat shock proteins increase resistance to apoptosis. Exp Cell Res 1996; 223(1):163-170.
36. Samali A, Orrenius S. Heat shock proteins: Regulators of stress response and apoptosis. Cell Stress Chaperones 1998; 3(4):228-236.
37. Concannon CG, Gorman AM, Samali A. On the role of Hsp27 in regulating apoptosis. Apoptosis 2003; 8(1):61-70.
38. Yoshida H, Kong YY, Yoshida R et al. Apafl is required for mitochondrial pathways of apoptosis and brain development. Cell 1998; 94(6):739-750.
39. Kuida K, Haydar TF, Kuan CY et al. Reduced apoptosis and cytochrome c-mediated caspase activation in mice lacking caspase 9. Cell 1998; 94(3):325-337.
40. Kuida K, Zheng TS, Na S et al. Decreased apoptosis in the brain and premature lethality in CPP32-deficient mice. Nature 1996; 384(6607):368-372.
41. Yuen EC, Howe CL, Li Y et al. Nerve growth factor and the neurotrophic factor hypothesis. Brain Dev 1996; 18(5):362-368.
42. Mearow KM, Dodge ME, Rahimtula M et al. Stress-mediated signaling in PC12 cells - the role of the small heat shock protein, Hsp27, and Akt in protecting cells from heat stress and nerve growth factor withdrawal. J Neurochem 2002; 83(2):452-462.
43. Caviness Jr VS, Takahashi T, Nowakowski RS. Numbers, time and neocortical neuronogenesis: A general developmental and evolutionary model. Trends Neurosci 1995; 18(9):379-383.
44. de la Rosa EJ, de Pablo F. Cell death in early neural development: Beyond the neurotrophic theory. Trends Neurosci 2000; 23(10):454-458.
45. Morales AV, Hadjiargyrou M, Diaz B et al. Heat shock proteins in retinal neurogenesis: Identification of the PM1 antigen as the chick Hsc70 and its expression in comparison to that of other chaperones. Eur J Neurosci 1998; 10(10):3237-3245.
46. de la Rosa EJ, Vega-Nunez E, Morales AV et al. Modulation of the chaperone heat shock cognate 70 by embryonic (pro)insulin correlates with prevention of apoptosis. Proc Natl Acad Sci USA 1998; 95(17):9950-9955.
47. Rubio E, Valenciano AI, Segundo C et al. Programmed cell death in the neurulating embryo is prevented by the chaperone heat shock cognate 70. Eur J Neurosci 2002; 15(10):1646-1654.
48. Mehlen P, Coronas V, Ljubic-Thibal V et al. Small stress protein Hsp27 accumulation during dopamine-mediated differentiation of rat olfactory neurons counteracts apoptosis. Cell Death Differ 1999; 6(3):227-233.
49. Liang P, MacRae TH. Molecular chaperones and the cytoskeleton. J Cell Sci 1997; 110(Pt 13):1431-1440.
50. da Silva JS, Dotti CG. Breaking the neuronal sphere: Regulation of the actin cytoskeleton in neuritogenesis. Nat Rev Neurosci 2002; 3(9):694-704.
51. Lariviere RC, Julien JP. Functions of intermediate filaments in neuronal development and disease. J Neurobiol 2004; 58(1):131-148.
52. Pekny M, Pekna M. Astrocyte intermediate filaments in CNS pathologies and regeneration. J Pathol 2004; 204(4):428-437.
53. Lavoie JN, Lambert H, Hickey E et al. Modulation of cellular thermoresistance and actin filament stability accompanies phosphorylation-induced changes in the oligomeric structure of heat shock protein 27. Mol Cell Biol 1995; 15(1):505-516.
54. Guay J, Lambert H, Gingras-Breton G et al. Regulation of actin filament dynamics by p38 map kinase-mediated phosphorylation of heat shock protein 27. J Cell Sci 1997; 110(Pt 3):357-368.
55. Miron T, Vancompernolle K, Vandekerckhove J et al. A 25-kD inhibitor of actin polymerization is a low molecular mass heat shock protein. J Cell Biol 1991; 114(2):255-261.
56. Lavoie JN, Hickey E, Weber LA et al. Modulation of actin microfilament dynamics and fluid phase pinocytosis by phosphorylation of heat shock protein 27. J Biol Chem 1993; 268(32):24210-24214.
57. Benndorf R, Hayess K, Ryazantsev S et al. Phosphorylation and supramolecular organization of murine small heat shock protein HSP25 abolish its actin polymerization-inhibiting activity. J Biol Chem 1994; 269(32):20780-20784.
58. Minowada G, Welch W. Variation in the expression and/or phosphorylation of the human low molecular weight stress protein during in vitro cell differentiation. J Biol Chem 1995; 270(13):7047-7054.
59. Costigan M, Mannion RJ, Kendall G et al. Heat shock protein 27: Developmental regulation and expression after peripheral nerve injury. J Neurosci 1998; 18(15):5891-5900.

60. Williams KL, Rahimtula M, Mearow KM. Hsp27 and axonal growth in adult sensory neurons in vitro. BMC Neurosci 2005; 6(1):24.
61. Davidson SM, Morange M. Hsp25 and the p38 MAPK pathway are involved in differentiation of cardiomyocytes. Dev Biol 2000; 218(2):146-160.
62. Hino M, Kurogi K, Okubo MA et al. Small heat shock protein 27 (HSP27) associates with tubulin/microtubules in HeLa cells. Biochem Biophys Res Commun 2000; 271(1):164-169.
63. Ahmad S, Ahuja R, Venner TJ et al. Identification of a protein altered in mutants resistant to microtubule inhibitors as a member of the major heat shock protein (hsp70) family. Mol Cell Biol 1990; 10(10):5160-5165.
64. Sanchez C, Padilla R, Paciucci R et al. Binding of heat-shock protein 70 (hsp70) to tubulin. Arch Biochem Biophys 1994; 310(2):428-432.
65. Sanchez ER, Redmond T, Scherrer LC et al. Evidence that the 90-kilodalton heat shock protein is associated with tubulin-containing complexes in L cell cytosol and in intact PtK cells. Mol Endocrinol 1988; 2(8):756-760.
66. Kirby BA, Merril CR, Ghanbari H et al. Heat shock proteins protect against stress-related phosphorylation of tau in neuronal PC12 cells that have acquired thermotolerance. J Neurosci 1994; 14(9):5687-5693.
67. Dou F, Netzer WJ, Tanemura K et al. Chaperones increase association of tau protein with microtubules. Proc Natl Acad Sci USA 2003; 100(2):721-726.
68. Perng MD, Cairns L, van den IP et al. Intermediate filament interactions can be altered by HSP27 and alphaB-crystallin. J Cell Sci 1999; 112(Pt 13):2099-2112.
69. Cleveland DW, Hoffman PN. Slow axonal transport models come full circle: Evidence that microtubule sliding mediates axon elongation and tubulin transport. Cell 1991; 67(3):453-456.
70. Green LA, Liem RK. Beta-internexin is a microtubule-associated protein identical to the 70-kDa heat-shock cognate protein and the clathrin uncoating ATPase. J Biol Chem 1989; 264(26):15210-15215.
71. Napolitano EW, Pachter JS, Chin SS et al. beta-Internexin, a ubiquitous intermediate filament-associated protein. J Cell Biol 1985; 101(4):1323-1331.
72. Kaplan DR, Miller FD. Neurotrophin signal transduction in the nervous system. Curr Opin Neurobiol 2000; 10(3):381-391.
73. Konishi H, Matsuzaki H, Tanaka M et al. Activation of protein kinase B (Akt/RAC-protein kinase) by cellular stress and its association with heat shock protein Hsp27. FEBS Lett 1997; 410(2-3):493-498.
74. Murashov AK, Ul Haq I, Hill C et al. Crosstalk between p38, Hsp25 and Akt in spinal motor neurons after sciatic nerve injury. Brain Res Mol Brain Res 2001; 93(2):199-208.
75. Rane MJ, Coxon PY, Powell DW et al. p38 Kinase-dependent MAPKAPK-2 activation functions as 3-phosphoinositide-dependent kinase-2 for Akt in human neutrophils. J Biol Chem 2001; 276(5):3517-3523.
76. Stokoe D, Engel K, Campbell DG et al. Identification of MAPKAP kinase 2 as a major enzyme responsible for the phosphorylation of the small mammalian heat shock proteins. FEBS Lett 1992; 313(3):307-313.
77. Sato S, Fujita N, Tsuruo T. Modulation of Akt kinase activity by binding to Hsp90. Proc Natl Acad Sci USA 2000; 97(20):10832-10837.
78. Rane MJ, Pan Y, Singh S et al. Heat shock protein 27 controls apoptosis by regulating Akt activation. J Biol Chem 2003; 278(30):27828-27835.
79. Yao R, Cooper GM. Requirement for phosphatidylinositol-3 kinase in the prevention of apoptosis by nerve growth factor. Science 1995; 267(5206):2003-2006.
80. Dudek H, Datta SR, Franke TF et al. Regulation of neuronal survival by the serine-threonine protein kinase Akt. Science 1997; 275(5300):661-665.
81. Philpott KL, McCarthy MJ, Klippel A et al. Activated phosphatidylinositol 3-kinase and Akt kinase promote survival of superior cervical neurons. J Cell Biol 1997; 139(3):809-815.
82. Crowder RJ, Freeman RS. Phosphatidylinositol 3-kinase and Akt protein kinase are necessary and sufficient for the survival of nerve growth factor-dependent sympathetic neurons. J Neurosci 1998; 18(8):2933-2943.
83. Eves EM, Xiong W, Bellacosa A et al. Akt, a target of phosphatidylinositol 3-kinase, inhibits apoptosis in a differentiating neuronal cell line. Mol Cell Biol 1998; 18(4):2143-2152.
84. Kim Y, Seger R, Suresh Babu CV et al. A positive role of the PI3-K/Akt signaling pathway in PC12 cell differentiation. Mol Cells 2004; 18(3):353-359.
85. Higuchi M, Onishi K, Masuyama N et al. The phosphatidylinositol-3 kinase (PI3K)-Akt pathway suppresses neurite branch formation in NGF-treated PC12 cells. Genes Cells 2003; 8(8):657-669.

86. Gao T, Newton AC. The turn motif is a phosphorylation switch that regulates the binding of Hsp70 to protein kinase C. J Biol Chem 2002; 277(35):31585-31592.
87. Patapoutian A, Reichardt LF. Trk receptors: Mediators of neurotrophin action. Curr Opin Neurobiol 2001; 11(3):272-280.
88. Pang L, Sawada T, Decker SJ et al. Inhibition of MAP kinase kinase blocks the differentiation of PC-12 cells induced by nerve growth factor. J Biol Chem 1995; 270(23):13585-13588.
89. Nobes CD, Tolkovsky AM. Neutralizing anti-p21ras Fabs suppress rat sympathetic neuron survival induced by NGF, LIF, CNTF and cAMP. Eur J Neurosci 1995; 7(2):344-350.
90. Cowley S, Paterson H, Kemp P et al. Activation of MAP kinase kinase is necessary and sufficient for PC12 differentiation and for transformation of NIH 3T3 cells. Cell 1994; 77(6):841-852.
91. Bijur GN, Jope RS. Opposing actions of phosphatidylinositol 3-kinase and glycogen synthase kinase-3beta in the regulation of HSF-1 activity. J Neurochem 2000; 75(6):2401-2408.
92. Xavier IJ, Mercier PA, McLoughlin CM et al. Glycogen synthase kinase 3beta negatively regulates both DNA-binding and transcriptional activities of heat shock factor 1. J Biol Chem 2000; 275(37):29147-29152.
93. Goldstein JL, Anderson RG, Brown MS. Coated pits, coated vesicles, and receptor-mediated endocytosis. Nature 1979; 279(5715):679-685.
94. Chappell TG, Welch WJ, Schlossman DM et al. Uncoating ATPase is a member of the 70 kilodalton family of stress proteins. Cell 1986; 45(1):3-13.
95. de Waegh S, Brady ST. Axonal transport of a clathrin uncoating ATPase (HSC70): A role for HSC70 in the modulation of coated vesicle assembly in vivo. J Neurosci Res 1989; 23(4):433-440.
96. Black MM, Chestnut MH, Pleasure IT et al. Stable clathrin: Uncoating protein (hsc70) complexes in intact neurons and their axonal transport. J Neurosci 1991; 11(5):1163-1172.
97. Tsang TC. New model for 70 kDa heat-shock proteins' potential mechanisms of function. FEBS Lett 1993; 323(1-2):1-3.
98. Giuffrida AM, Gadaleta MN, Serra I et al. Mitochondrial DNA, RNA, and protein synthesis in different regions of developing rat brain. Neurochem Res 1979; 4(1):37-52.
99. Renis M, Cantatore P, Loguercio Polosa P et al. Content of mitochondrial DNA and of three mitochondrial RNAs in developing and adult rat cerebellum. J Neurochem 1989; 52(3):750-754.

CHAPTER 4

Heme Oxygenase as a Therapeutic Funnel in Nutritional Redox Homeostasis and Cellular Stress Response: Role of Acetylcarnitine

Vittorio Calabrese,* Giovanni Pennisi, Menotti Calvani, D. Allan Butterfield, Cesare Mancuso and Anna Maria Giuffrida Stella

Abstract

Reduction of cellular expression and activity of antioxidant proteins and the consequent increase of oxidative stress are fundamental causes for both the aging processes and neurodegenerative diseases. Oxidative stress has been implicated in mechanisms leading to neuronal cell injury in various pathological states of the brain. Alzheimer's disease (AD) is a progressive disorder with cognitive and memory decline, speech loss, personality changes and synapse loss. Many approaches have been undertaken to understand AD, but the heterogeneity of the etiologic factors makes it difficult to define the clinically most important factor determining the onset and progression of the disease. There is now evidence to suggest that networks of responses exist in the brain to detect and control diverse forms of stress. This is accomplished by a complex network of the so-called longevity assurance processes, which are composed of several genes termed *vitagenes*. Among these, heat shock proteins form a highly conserved system responsible for the preservation and repair of the correct protein conformation. Recent studies have shown that the heat shock response contributes to establish a cytoprotective state in a wide variety of human diseases, including inflammation, cancer, aging and neurodegenerative disorders. Given the broad cytoprotective properties of the heat shock response there is now strong interest in discovering and developing pharmacological agents capable of inducing the heat shock response. Acetylcarnitine (LAC) is proposed as a therapeutic agent for several neurodegenerative disorders, and there is evidence that LAC may play a critical role as a modulator of cellular stress response in health and disease states. In the present review we discuss the role of the heme oxygenase pathway in cellular stress response. We then review the evidence for the role of acetylcarnitine in modulating redox-dependent mechanisms leading to up-regulation of vitagenes in brain, and hence potentiate brain stress tolerance.

Introduction

It is well established that living cells are constantly challenged by conditions which cause acute or chronic stress. The brain has a large potential oxidative capacity but a limited ability to

*Corresponding Author: Vittorio Calabrese—Department of Chemistry, Biochemistry and Molecular Biology Section, Faculty of Medicine, University of Catania, Catania, Italy. Email: calabres@mbox.unict.it

Heat Shock Proteins in Neural Cells, edited by Christiane Richter-Landsberg.

counteract oxidative stress.[1-3] Within the cell, reactive oxygen species (ROS) are physiologically present at minimal concentration as by-products of aerobic metabolism as well as second messengers in many signal transduction pathways and, in normal conditions, there is a steady-state balance between pro-oxidants and antioxidants which is necessary to ensure optimal efficiency of antioxidant defenses.[4-7] However, when the rate of free radical generation exceeds the capacity of antioxidant defenses, oxidative stress ensues with consequential severe damage to DNA, protein and lipids.[8-10] Oxidative stress has been implicated in mechanisms leading to neuronal cell injury in various pathological states of the brain, including neurodegenerative disorders such as Alzheimer's disease (AD).[11-15]

Recently the term "nitrosative stress" has been used to indicate the cellular damage elicited by nitric oxide and its congeners peroxynitrite, N_2O_3, nitroxyl anion and nitrosonium (all can be indicated as reactive nitrogen species or RNS).[16-18]

From a molecular point of view, the cell is able to fight against oxidant stress using many resources, including vitamins (A, C and E), bioactive molecules (glutathione, thioredoxin, flavonoids), enzymes (Heat shock protein-32, superoxide dismutase, catalase, glutathione peroxidases, thioredoxin reductase, etc) and redox sensitive protein transcriptional factors (AP-1, NFkB, Nrf-2, HSF, etc). The heat shock proteins (Hsps) are one of the more studied defense system active against cellular damage.

In this chapter we describe the more recent discoveries about the biochemical changes occurring in the central nervous system (CNS), when brain cells are exposed to chronic oxidative insult as well as the key role played by the heat shock response, particularly the heme oxygenase (Hsp32) and Hsp70 pathways, in modulating the onset and progression of AD. Whether or not stress proteins are neuroprotective is still under debate; however, emerging evidence underscores the high potential of the Hsp system as a target for new neuroprotective strategies, expecially those aimed at minimizing deleterious consequences associated with oxidative stress, such as in neurodegenerative disorders and brain aging. We review here also the evidence for the role of acetylcarnitine in modulating redox-dependent mechanisms leading to up-regulation of vitagenes in brain, and hence potentiate brain stress tolerance.

Heme Oxygenase-1

Heme oxygenase-1 (HO-1), also referred to as heat shock protein-32, is the redox-sensitive inducible isoform of the HO family. Heme oxygenase is a microsomal enzyme and catalyzes the degradation of heme in a multistep, energy-requiring system. The reaction catalyzed by HO is the α-specific oxidative cleavage of heme moieties to form equimolar amounts of ferrous iron, carbon monoxide (CO) and biliverdin. This latter is then reduced by the cytosolic enzyme biliverdin reductase to bilirubin (BR), which is then conjugated with glucuronic acid and excreted.[19]

Increasing evidence suggested that the HO-1 gene is redox regulated (Fig. 1) and contains in its promoter region the antioxidant responsive element (ARE), similar to other antioxidant enzymes.[20] In fact HO-1 can be induced by several stimuli including oxidative and nitrosative stress, ischemia, heat shock, LPS, hemin and the neuroprotective agent neotrofin.[21-23] HO-1 induction is one of the earlier cellular responses to tissue damage and is responsible for the rapid transformation of the pro-oxidant heme into CO and BR, two molecules with anti-inflammatory and anti-oxidant activity.[24-27]

The HO-1 gene is induced by other factors (Fig. 2), including metallophorphyrins and hemin, as well as ultraviolet A (UVA) irradiation, hydrogen peroxide, pro-oxidant states or inflammation.[28,29] This characteristic inducibility of HO-1 gene strictly relies on its configuration: the 6.8-kilobase gene is organized into 4 introns and 5 exons. A promoter sequence is located approximately 28 base pairs upstream from the transcriptional site of initiation. In addition, different transcriptional enhancer elements, such as heat shock element and metal regulatory element reside in the flanking 5' region. Also, inducer-responsive sequences have been identified in the proximal enhancer located upstream the promoter and, more distally, in two enhancers located 4kb and 10 kb upstream the initiation site.[30] The molecular mechanism

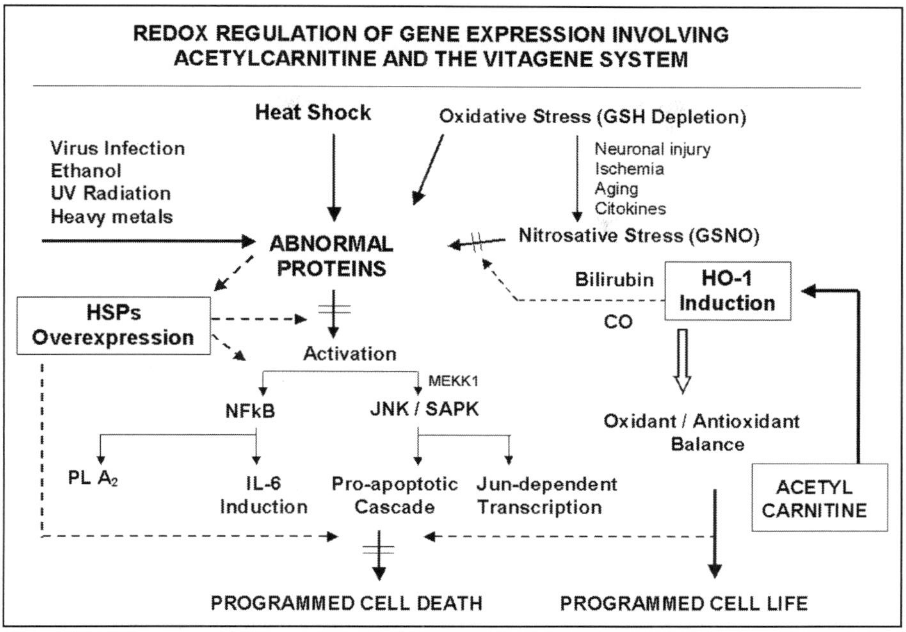

Figure 1. Redox regulation of gene expression involving Acetylcarnitine and the Vitagene system. Proposed role for acetylcarnitine and the *vitagene* member HSPs, in modulation cellular redox state and cell stress tolerance. Various proteotoxic (or genotoxic) conditions cause depletion of free HSPs that lead to activation of stress kinase and proinflammatory and apoptotic signaling pathways. HSP70 prevents stress-induced apoptosis by interfering with the SAPK/JNK signaling and by blocking caspase proteolytic cascade. Nitrosative-dependent thiol depletion triggers HO-1 induction, and increased HO-1 activity is translated into augmented production of carbon monoxide and the antioxidant bilirubin. Exogenous non toxic inducers, such as acetylcarnitine or polyphenols can counteract increased NOS activity and NO-mediated cytotoxicity through up-regulation of the HO system. HO-1 may directly decrease NO synthase protein levels by degrading the cofactor heme. (PLA_2: phospholipase A_2; IL-6: interleukin-6; AP-1: activator protein-1; SAPK: stress-activated protein kinase; JNK: c-jun N-terminal kinase; NFκB: nuclear factor kappa-B; GSNO: S-nitrosoglutathione; HO-1: heme oxygenase-1).

that confers inducible expression of *ho-1* in response to numerous and diverse conditions has remained elusive. One important clue has recently emerged from a detailed analysis of the transcriptional regulatory mechanisms controlling the mouse and human *ho-1* genes. The induction of *ho-1* is regulated principally by two upstream enhancers, E1 and E2.[31] Both enhancer regions contain multiple stress (or antioxidant) responsive elements (StRE, also called ARE) that also conform to the sequence of the Maf recognition element (MARE)[32] with a consensus sequence (GCnnnGTA) similar to that of other antioxidant enzymes.[33] There is evidence to suggest that heterodimers of NF-E2-related factors 2 (Nrf2) and one or another of the small Maf proteins (i.e., MafK, mafF and MafG) are directly involved in induction of *ho-1* through these MAREs.[32] A possible model, centered on Nrf2 activity, suggests that the *ho-1* locus is situated in a chromatin environment that is permissive for activation. Since the MARE can be bound by various heterodimeric basic leucine zipper (bZip) factors including NF-E2, as well as several other NF-E2-related factors (Nrf1, Nrf2, and Nrf3), Bach, Maf and AP-1 families,[31] random interaction of activators with the *ho-1* enhancers would be expected to cause spurious expression. This raises a paradox as to how cells reduce transcriptional noise from the *ho-1* locus in the absence of metabolic or environmental stimulation. This problem could be

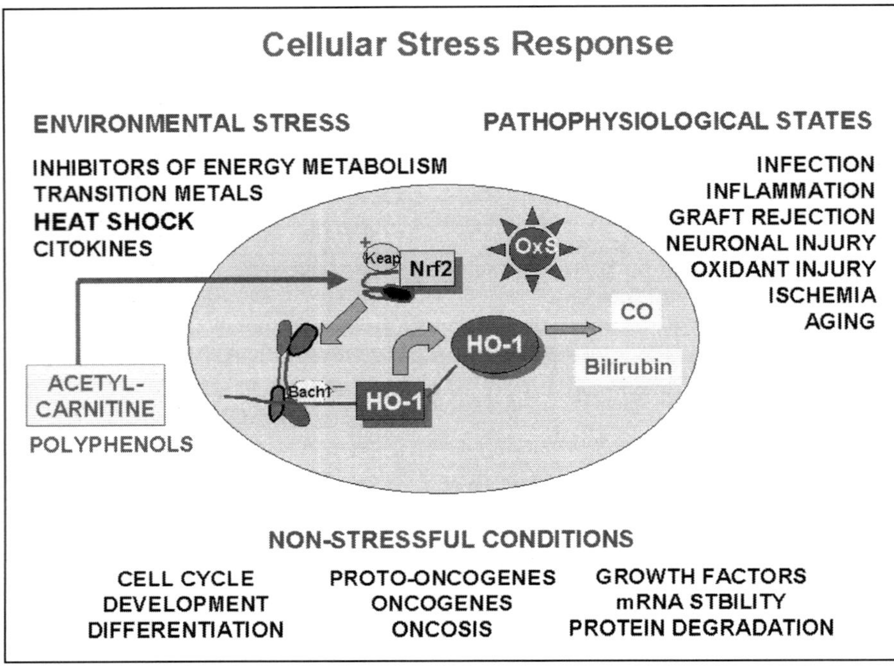

Figure 2. Physiological and pathophysiological conditions inducing cellular stress response. Environmental stress factors, such as heavy metals, cytokines, heat shock or energy metabolism inhibitors or pathophysiological conditions of oxidant antioxidant balance perturbation such as inflammation, graft rejections, neuronal damage, ischemia and brain aging are all situations associated with induction of cellular stress response. Hsp response is also involved in cellular homeostasis during various physiological conditions, such as during brain development and differentiation, cell cycle, apoptosis and oncosis, oncogene and growth factors action, as well as mRNA and protein half-life. The heme oxygenase system represents a therapeutic funnel for cellular stress tolerance and can be activated by non noxious stimuli, such as nutritional antioxidants or acetylcarnitine. Acetylcarnitine, through activation of the redox sensitive transcription factor Nrf-2, by up-regulating HO-1 may counteract nitrosative stress and NO-mediated neurotoxicity. In the same figure are described the respective roles of protein factors Bach-2 (positive) and Keap (negative) in Nrf2 activation.

reconciled by the activity of repressors that prevent nonspecific activation. One possible candidate is the heme protein Bach1, a transcriptional repressor endowed with DNA binding activity, which is negatively regulated upon binding with heme. Bach1-heme interaction is mediated by evolutionarily conserved heme regulatory motifs (HRM), including the cysteine-proline dipeptide sequence in Bach1. Hence, a plausible model accounting for the regulation of *ho-1* expression by Bach1 and heme, is that expression of *ho-1* gene is regulated through antagonism between transcription activators and the repressor Bach1 (Fig. 2). While under normal physiological conditions expression of *ho-1* is repressed by Bach1/Maf complex, increased levels of heme displace Bach1 from the enhancers and allow activators, such as heterodimer of Maf or Keap with NF-E2 related activators (Nrf2), to the transcriptional promotion of *ho-1* gene[31] (Fig. 2). To our knowledge, the Bach1- *ho-1* system is the first example in higher eukaryotes that involves a direct regulation of a transcription factor for an enzyme gene by its substrate. Thus, regulation of *ho-1* involves a direct sensing of heme levels by Bach1 (by analogy to *lac* repressor sensitivity to lactose), generating a simple feedback loop whereby the substrate affects repressor-activator antagonism.

The promoter region also contains two metal responsive elements, similar to those found in metallothionein-1 gene, which respond to heavy metals (cadmium and zinc) only after recruitment of another fragment located upstream, between -3.5 and 12 kbp (CdRE). In addition, a 163-bp fragment containing two binding sites for HSF-1, which mediates the HO-1 transcription are located 9.5 kb upstream of the initiation site.[34] The distal enhancer regions are important in regulating HO-1 in inflammation, since, as has been demonstrated, they are responsive to endotoxin. In the promoter region also resides a 56 bp fragment which responds to the STAT-3 acute-phase response factor, involved in the down-regulation of HO-1 gene induced by glucocorticoid.[35,36]

HO-1, Oxidative Stress and Neurodegenerative Disorders

The mechanisms responsible for neuronal death are not completely elucidated, even if many studies suggest that ROS are primarily involved in the genesis of neurodegenerative disorders.[11-15,37-39] Due to its strong antioxidant properties and wide distribution within the CNS HO-1 has been proposed as a key enzyme in the prevention of brain damage.[21,22,40] Recently, Panahian et al using transgenic mice over-expressing HO-1 in neurons, demonstrated the neuroprotective effect of this enzyme in an experimental model of ischemic brain damage.[41] The neuroprotective effects of over-expressed HO-1 can be attributed to: (i) increase in cGMP and bcl-2 levels in neurons; (ii) inactivation of p53, a protein involved in promoting cell death; (iii) increase in antioxidant sources and (iv) increase in the iron sequestering protein, ferritin.[41] Particularly interesting is the role played by HO-1 in AD, a neurodegenerative disorder which involves a chronic inflammatory response associated with both oxidative brain injury and β-amyloid associated pathology. Significant increases in the levels of HO-1 have been observed in AD brains in association with neurofibrillary tangles and also HO-1 mRNA was found increased in AD neocortex and cerebral vessels.[42,43] The HO-1 increase was not only in association with neurofibrillary tangles, but also colocalized with senile plaques and glial fibrillary acidic protein-positive astrocytes in AD brains.[44] It is plausible that the dramatic increase in HO-1 in AD may be a direct response to an increase in free heme concentrations, associated with neurodegeneration, and can be considered as an attempt of brain cells to convert the highly toxic heme into the antioxidants CO and BR.

The protective role played by HO-1 and its products in AD raised new possibilities regarding the possible use of natural substances, which are able to increase HO-1 levels, as potential drugs for the treatment of AD. In this light, very promising are the polyphenolic compounds contained in some herbs and spices, e.g., curcumin.[45-47] Curcumin is the active anti-oxidant principle in *Curcuma longa*, a colouring agent and food additive commonly used in Indian culinary preparations. This polyphenolic substance has the potential to inhibit lipid peroxidation and to effectively intercept and neutralize ROS and RNS.[48] In addition, curcumin has been shown to significantly increase HO-1 in astrocytes and vascular endothelial cells.[46,49] This latter effect on HO-1 can explain, at least in part, the anti-oxidant properties of curcumin, in particular keeping in mind that HO-1-derived BR has the ability to scavenge both ROS and RNS.[24,27,50,51] Epidemiological studies suggested that curcumin, as one of the most prevalent nutritional and medicinal compounds used by the Indian population, is responsible for the significantly reduced (4.4- fold) prevalence of AD in India compared to United States.[52] Based on these findings, Lim and colleagues have provided convincing evidence that dietary curcumin given to an Alzheimer transgenic APPSw mouse model (Tg2576) for 6 months resulted in a suppression of indices of inflammation and oxidative damage in the brain of these mice.[53] Furthermore, in a human neuroblastoma cell line it has recently been shown that curcumin inhibits NFkB activation, efficiently preventing neuronal cell death.[48]

Although it is generally agreed that HO-1 over-expression is a common feature during oxidative stress, recent papers demonstrated that HO-1 can be repressed following oxidant conditions. In particular human cells exposed to hypoxia, thermal stress and interferon-γ treatment showed a marked HO-1 repression and this effect seems to be peculiar for human, because rodent cells over-expressed HO-1 when exposed to the same stimuli.[54-57] The importance of HO-1 repression has been corroborated by the discovery of Bach-1/Bach-2 as heme-regulated transcription factors for the HO-1 gene.[58] In fact, Bach-1 is broadly expressed in mice and human tissues and, in human cells, it is induced by the same stimuli which are able to repress the HO-1 gene.[54,59-61] The reason why the cell should react to an oxidant stress by repressing HO-1 gene is strictly related to the maintenance of a good metabolic balance during stressful conditions. The current hypothesis suggests that HO-1 repression is useful for the cell because this (i) decreases the energy costs necessary for heme degradation; (ii) reduces the accumulation of CO and BR, which can become toxic if produced in excess; and (iii) increases the intracellular content of heme necessary for the preservation of vital functions such as respiration and defense.[60]

Carbon Monoxide and Stress Response

Carbon monoxide (CO) is the gaseous products of HO and it has been found to play a role in several biological phenomena, including hippocampal long-term potentiation, nonadrenergic noncholinergic gastrointestinal relaxation and vasodilatation, and is currently regarded as a neuromodulator in the peripheral and central nervous system (for an extensive review on CO and its functions in the nervous system see ref. 40). Evidence from in vitro and in vivo studies suggests that the HO-CO pathway is involved in the modulation of the neuroendocrine mechanism of stress. Thus, increased CO generation is clearly associated with the inhibition of K^+ stimulated arginine vasopressin (AVP) and oxytocin release from rat hypothalamic explants, whereas the inhibition of HO activity significantly potentiates the LPS-induced increase in AVP circulating levels while reducing the hypothalamic content of this neuropeptide.[62-64] With regards to corticotropin-releasing hormone (CRH), the effects of CO on the release of this hormone are contradictory, since increases in CO generation induced by the HO substrates, hematin and hemin, were associated with reduced or enhanced CRH release respectively, in two different in vitro models.[65,66] As far as the intracellular mechanism(s) by which CO exerts its biological functions, it is generally agreed that this gas activates the cytosolic form of guanylyl cyclase (sGC), which in turn increases intracellular cGMP levels.[22] However during the last ten years many studies have appeared in literature demonstrating that CO signals through the activation of alternative intracellular signal transduction pathways. Studies from our laboratory suggested that the activation of another hemoprotein, cyclooxygenase (COX), plays a significant role in CO signaling in the rat hypothalamus. In these studies we demonstrated that hemin, the precursor of CO via HO, dose-dependently increases PGE_2 production from rat hypothalamus in vitro and this effect is specifically due to CO because it is counteracted by the HO inhibitor Sn-mesoporphyrin-IX and oxyhemoglobin, the latter being a well known scavenger for CO.[67] The direct evidence about the stimulatory role of CO on PGs production was obtained by incubating hypothalami directly in CO saturated solutions and measuring significantly increased PGE_2 levels with respect to control tissue.[25] Recently Jaggar et al,[68] in a very elegant paper, demonstrated that exogenous or endogenously produced CO dilates cerebral arterioles by directly activating large-conductance Ca^{2+}-activated K^+ (K_{Ca}) channels primarily by increasing the coupling ratio and amplitude relationship between Ca^{2+} sparks and K_{Ca} channels. Although CO is a potent and effective activator of K_{Ca} channels, the gas does not dilate arterioles in the absence of Ca^{2+} sparks. Therefore, CO appears to act by priming K_{Ca} channels for activation by Ca^{2+} sparks, and this ultimately leads to arteriole dilation via membrane hyperpolarization.[68] Finally, Otterbein et al[26] have shown that in organs and tissues different from brain, CO exerts anti-inflammatory and anti-apoptotic effects dependent on the modulation of the p38 MAPK-signaling pathway. By virtue of these effects, CO confers protection in oxidative lung injury models, and likely plays a role in HO-1 mediated tissue protection.[69]

Heat Shock Protein-70

The 70 kDa family of stress proteins is one of the most extensively studied. Included in this family are Hsc70 (heat shock cognate, the constitutive form), Hsp70 (the inducible form, also referred to as Hsp72) and GRP-75 (a constitutively expressed glucose-regulated protein found in the endoplasmic reticulum).

Only recently, the availability of transgenic animals and gene transfer allowed to over-express the gene encoding for Hsp70, thus demonstrating that overproduction of this protein leads to protection in several different models of nervous system injury.[70,71] Following focal cerebral ischemia, Hsp70 mRNA is synthesized in most ischemic cells except in areas of very low blood flow, due to scarce ATP levels. Hsp70 proteins are produced mainly in endothelial cells, in the core of infarcts in the cells that are most resistant to ischemia, in glial cells at the edges of infarcts and in neurons outside the areas of infarction[72] It has been suggested that this neuronal expression of Hsp70 outside an infarct can be used to define the ischemic penumbras, i.e., the zone of protein denaturation in the ischemic areas.[72]

As mentioned above, Hsps are induced in many neurodegenerative disorders mainly in the view of its cytoprotective function. Hsp72 was overexpressed in post-mortem cortical tissue of AD patients and an increase in Hsp70 mRNA was found in cerebellum, hippocampus and cortex of AD patients during the agonal phase of the disease.[73-75] Recently Kakimura et al[76] demonstrated that Hsp70 induces IL-6 and TNF-α in microglial cells, and this event is associated with an increased phagocytosis and clearance of Aβ peptides. The same authors hypothesize that Hsps could activate microglial cells through NFkB and p-38 MAPK-dependent pathways.[76]

A large body of evidence now suggest a correlation between mechanisms of nitrosative stress and Hsp induction. We have demonstrated in astroglial cell cultures that cytokine-induced nitrosative stress is associated with an increased synthesis of Hsp70 stress proteins. The molecular mechanisms regulating the NO-induced activation of heat-shock signal seems to involve cellular oxidant/antioxidant balance, mainly represented by the glutathione status and the antioxidant enzymes.[77,78]

Acetylcarnitine

Mitochondria are cellular organelles involved in many metabolic processes such as pyruvate oxidation, the tricarboxylic acid cycle, fatty acid β-oxidation and are the common final pathway of oxidative phosphorylation, which generates most of the cellular energetic source, ATP. It has been proposed that accumulation of mitochondrial DNA (mtDNA) during life is a major cause of age-related disease and this is because of its high mutagenic propensity. The lack of introns and protective histones, limited nucleotide excision and recombination DNA repair mechanisms, location in proximity of the inner mitochondrial membrane which exposes mtDNA to an enriched free radical milieu, are all factors contributing to a 10-fold higher mutation rate occurring in the mtDNA than in the nuclear DNA. Relevant to mitochondrial bioenergetics, in fact, is the finding of a significant decrease in state 3/state 4 ratio, which has been observed to occur in brain as function of age.[79] Since this respiratory control ratio relates to the coupling efficiency between electron flux through the electron transport chain and ATP production, an increase in state 4 would result in a more reductive state of mitochondrial complexes and, consequently, to an increase in free radical species production. A decrease in state 3/state 4 respiration during aging has been found associated with a significant decrease in cardiolipin content in brain mitochondria.[80] This loss could play a critically important role in the age-related decrements in mitochondrial function, and appears to be associated with both quantitative and qualitative region-specific protein changes, which are parallel to structural changes, such as decrease of the inner membrane surface, smaller as well as sparser cristae, decreased fluidity and increased fragility. Modifications in cardiolipin composition are recognized to accompany functional changes in brain mitochondria, which include all proteins of the inner mitochondrial membrane that generally require interaction with cardiolipin for optimal catalytic activity.[81]

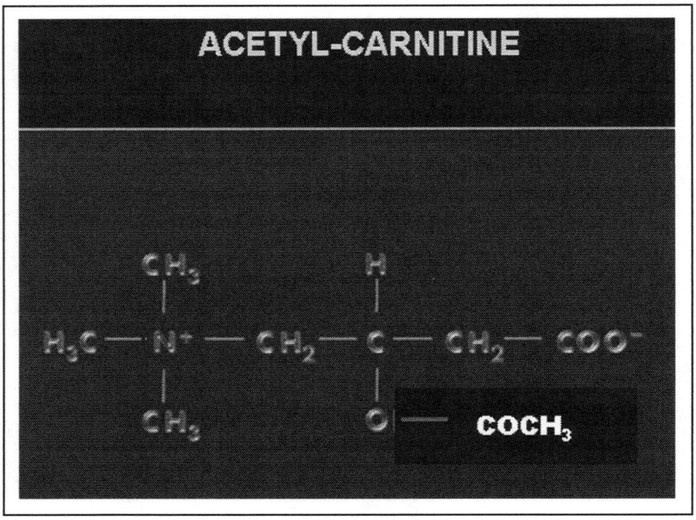

Figure 3. Chemical structure of Acetyl-L-carnitine.

Acetylcarnitine (LAC) (Fig. 3) is an ester of the trimethylated amino acid, L-carnitine, and is synthesised in the brain, liver, and kidney by the enzyme LAC-transferase. LAC facilitates the uptake of acetyl-CoA into the mitochondria during fatty acid oxidation, enhances acetylcholine production, and stimulates protein and membrane phospholipid synthesis.[82] At present, studies have shown that LAC is a compound of great interest for its wide clinical application in various neurological disorders: it may be of benefit in treating Alzheimer's dementia, chronic fatigue syndrome, depression in the elderly, HIV infection, diabetic neuropathies, ischemia and reperfusion of the brain, cognitive impairment of alcoholism, aging.[83-85] The neuroprotective benefits of this compound have been observed in the hippocampus, prefrontal cortex, substantia nigra and muscarinic receptor portions of the brain.[86] These benefits include antioxidant activity, improved mitochondrial energetics, stabilization of intracellular membranes and cholinergic neurotransmission.[87] Promising therapeutic applications of LAC are derived from observations that this compound crosses the blood-brain barrier through a saturable process in a sodium-dependent manner and improves neuronal energetic and repair mechanisms, while modifying acetylcholine production in the CNS.[88] LAC treatment restores the altered neurochemical abnormalities, cerebral energy metabolites in ischemia and aging and, in particular, ammonia-induced cerebral energy depletion.[87] In addition, it increases the responsiveness of aged neurons to neurotrophic factors in the CNS and it has preventive and corrective effects on diabetic neuropathology. Its beneficial effects have been also observed on EEG, evoked potentials and long-term synaptic potentiation in aged animals.[89] Moreover, LAC is commonly used also for the treatment of painful neuropathies: it exerts a potent analgesic effect by up-regulating metabotropic glutamate receptors.[90] There are experimental data that LAC improves memory function in Alzheimer's patients and it influences attention, learning and memory in the rat.[91] Chronic treatment enhances spatial acquisition in a novel environment of rats with behavioral impairments and has a slight effect on retention of the spatial discrimination in a familiar environment.[92] More recently, it has been observed that LAC produces sustained changes of nonassociative learning of sensitization and dyshabituation type in the invertebrate *Hirudo medicinalis*, and it has been suggested that LAC might exerts its effects by means of new protein synthesis, through qualitative and quantitative changes of gene expression. Furthermore, recent evidences have reported that LAC influences expression of glyoxylase 1, a gene involved in the detoxification of metabolic by-products, and increases p75-mRNA in Alzheimer's disease mutant transgenic mouse model Tg2576.[93] Recently, by using suppressive subtractive hybridisation (SSH) strategy, a PCR-based cDNA subtraction procedure particularly efficient for obtaining expressed transcripts often obscured by more abundant ones, it was reported that LAC modulates specific genes in the rat CNS, such as the hsp72 gene, the gene for the isoform of 14-3-3 protein and that encoding for the precursor mitochondrial P3 of ATP synthase lipid-binding protein.[94]

Acetylcarnitine fed to old rats increased cardiolipin levels to that of young rats and also restored protein synthesis in the inner mitochondrial membrane, as well as cellular oxidant/antioxidant balance, suggesting that administration of this compound may improve cellular bioenergetics in aged rats.[95] Fascinatingly, caloric restriction, a dietary regimen that extends life-span in rodents, maintains the levels of 18:2 acyl side chains and inhibits the cardiolipin composition changes.[96] In addition, caloric restriction was shown to retard the aging associated changes in oxidative damage, mitochondrial oxidant generation and antioxidant defenses observed during aging.[97,98]

Interestingly, we have recently demonstrated that acetylcarnitine treatment of astrocytes induces HO-1 in a dose and time dependent manner and that this effect was associated with up-regulation of other Hsps as well as high expression of the redox-sensitive transcription factor Nrf2 in the nuclear fraction of treated cells.[82] In addition, we showed that addition of LAC to astrocytes, prior to LPS and INFγ-induced nitrosative stress, prevents changes in mitochondrial respiratory chain complex activity, protein nitrosation and antioxidant status induced by inflammatory cytokine insult.[82] Very importantly, this new envisioned role of LAC as a molecule endowed with the capability of potentating the cellular stress response pathways appear to provide an alternative therapeutic approach for those pathophysiological conditions where stimulation of the HO pathway is warranted.[7] Although clinical application of compounds potentiating the action of stress responsive genes should be fully considered, a better understanding of how HO mediates its action will guide therapeutic strategies to enhance or suppress HO effects. Remarkably, the recent envisioned role of Hsp70 as a vehicle for intra-cytoplasmic and intra-nuclear delivery of fusion proteins or DNA to modulate gene expression[99,100] along with the evidence that binding of HO protein to HO-1 DNA modifies HO expression via nonenzymatic signaling events associated to CO and p38-dependent induction of Hsp70,[101] open intriguing perspectives, as it is possible to speculate that synergy between these two systems might represent a possible important target for the acetylcarnitine action, with possible impact on cell survival during times of oxidative stress, hence contributing to activation of cell life programs and to the extent of cellular stress tolerance.

Conclusions and Perspectives

Modulation of endogenous cellular defense mechanisms *via* the stress response signaling represents an innovative approach to therapeutic intervention in diseases causing tissue damage, such as neurodegeneration. Efficient functioning of maintenance and repair processes seems to be crucial for both survival and physical quality of life. This is accomplished by a complex network of the so-called longevity assurance processes, which are composed of several genes termed *vitagenes*. Consistently, by maintaining or recovering the activity of vitagenes it can be possible to delay the aging process and decrease the occurrence of age-related diseases with resulting prolongation of a healthy life span.[6,7,61] As one of the most important neurodegenerative disorders, AD is a progressive disorder with cognitive and memory decline, speech loss, personality changes and synapse loss. With the increasingly aging population of the United States, the number of AD patients is predicted to reach 14 million in the mid-21st century in the absence of effective interventions.[2,45] This will pose an immense economic and personal burden on the people of the U.S.A. Similar considerations apply worldwide, except in sub-Sahara Africa, where HIV infection rates seem to be leading to decreased incidence of AD. There is now strong evidence to suggest that factors such as oxidative stress and disturbed protein metabolism and their interaction in a vicious cycle are central to AD pathogenesis. Brain-accessible antioxidants, potentially, may provide the means of implementing this therapeutic strategy of delaying the onset of AD, and more in general all degenerative diseases associated with oxidative stress.[47,102] As one potentially successful approach, potentiation of endogenous secondary antioxidant systems can be achieved by interventions which target the HO-1/CO and/or Hsp70 systems. In this review, the importance of the stress response signaling and, in particular, the central role of HO-1 together with the redox-dependent mechanisms involved in cytoprotection are outlined. The beneficial effects of HO-1 induction result from heme degradation and cytoprotective regulatory functions of biliverdin/bilirubin redox cycling. Thus, HO-1 can amplify intracellular cytoprotective mechanisms against a variety of insults. Consequently, induction of HO-1, by increasing CO and/or biliverdin availability can be of clinical relevance.

Very importantly, HO-1 and CO can suppress the development of atherosclerotic lesions associated with chronic rejection of transplanted organs.[103] Consistently, LAC, as a molecule endowed with the capability of potentating the cellular stress response pathways, appears to afford similar protective action, thereby providing an alternative therapeutic approach valuable for all those pathophysiological conditions where stimulation of the HO pathway becomes a primary target.

Presented here is strong evidence that a crosstalk between stress response genes is critical for cell stress tolerance, highlighting compelling reasons for a renewed effort to understand the central role of this most extraordinary defense system in biology and medicine. All of the above evidence also supports the notion that stimulation of various maintenance and repair pathways through exogenous intervention, such as mild stress or compounds targeting the heat shock signal pathway, such as LAC, may have biological significance as a novel approach to delay the onset of various age-associated alterations in cells, tissues and organisms. Hence, by maintaining or recovering the activity of vitagenes it can be possible to delay the aging process and decrease the occurrence of age-related diseases with resulting prolongation of a healthy life span.

References

1. Calabrese V, Scapagnini G, Latteri S et al. Heme oxygenase, carbon monoxide and Cellular Stress Response in the Nervous System: The good and the enigmatic. Neurochemistry News 2002c; 2:57-60.
2. Calabrese V, Scapagnini G, Colombrita C et al. Redox regulation of heat shock protein expression in aging and neurodegenerative disorders associated with oxidative stress: A nutritional approach. Amino Acids 2003a; 25:437-444.
3. Poon HF, Calabrese V, Scapagnini G et al. Free radicals: Key to brain aging and heme oxygenase as a cellular response to oxidative stress. J Gerontology 2004a; 59:478-493.
4. Forman HJ, Fukuto JM, Torres M. Redox signaling: Thiol chemistry defines which reactive oxygen and nitrogen species can act as second messengers. Am J Physiol Cell Physiol 2004; 287:C246-C256.
5. Poon HF, Calabrese V, Scapagnini G et al. Free radicals and brain aging. Clin Geriatr Med 2004b; 20:329-359.
6. Calabrese V, Scapagnini G, Ravagna A et al. Increased expression of heat shock proteins in rat brain during aging: Relationship with mitochondrial function and glutathione redox state. Mech Age Dev 2004a; 125:325-335.
7. Calabrese V, Giuffrida Stella AM, Butterfield DA et al. Redox regulation in neurodegeneration and longevity: Role of the Heme Oxygenase and HSP70 systems in brain stress tolerance. Antioxid Redox Signal 2004b; 6:895-913.
8. Halliwell B. Hypothesis: Proteasomal dysfunction: A primary event in neurogeneration that leads to nitrative and oxidative stress and subsequent cell death. Ann NY Acad Sci 2002; 962:182-194.
9. Martindale JL, Holbrook NJ. Cellular response to oxidative stress: Signaling for suicide and survival. J Cell Physiol 2002; 192:1-15.
10. Bergamini CM, Gambetti S, Dondi A et al. Oxygen, reactive oxygen species and tissue damage. Curr Pharm Des 2004; 10:1611-1626.

11. Pappolla MA, Chyan YJ, Omar RA et al. Evidence of oxidative stress and in vivo neurotoxicity of beta-amyloid in a transgenic mouse model of Alzheimer's disease: A chronic oxidative paradigm for testing antioxidant therapies in vivo. Am J Pathol 1998; 152:871-877.
12. Smith MA, Hirai K, Hsiao K et al. Amyloid-beta deposition in Alzheimer transgenic mice is associated with oxidative stress. J Neurochem 1998; 70:2212-2215.
13. Butterfield DA, Drake J, Pocernich C et al. Evidence of oxidative damage in Alzheimer's disease brain: Central role for amyloid beta-peptide. Trends Mol Med 2001; 7:548-554.
14. Butterfield DA, Lauderback CM. Lipid peroxidation and protein oxidation in Alzheimer's disease brain: Potential causes and consequences involving amyloid β-peptide-associated free radical oxidative stress. Free Rad Biol Med 2002; 32:1050-1060.
15. Mattson MP. Pathways towards and away from Alzheimer's disease. Nature 2004; 430:631-639.
16. Drew B, Leeuwenburgh C. Aging and the role of reactive nitrogen species. Ann NY Acad Sci 2002; 959:66-81.
17. Kroncke KD. Nitrosative stress and transcription. Biol Chem 2003; 384:1365-1377.
18. Ridnour LA, Thomas DD, Mancardi D et al. The chemistry of nitrosative stress induced by nitric oxide and reactive nitrogen oxide species. Putting perspective on stressful biological situations. Biol Chem 2004; 385:1-10.
19. Maines MD. Heme oxygenase: Function, multiplicity, regulatory mechanisms, and clinical applications. FASEB J 1988; 2:2557-2568.
20. Prestera T, Talalay P, Alam J et al. Parallel induction of heme oxygenase-1 and chemoprotective phase 2 enzymes by electrophiles and antioxidants: Regulation by upstream antioxidant-responsive elements (ARE). Mol Med 1995; 1:827-837.
21. Maines MD. Heme Oxygenase in Clinical Applications and Functions. Boca Raton: CRC Press, 1992.
22. Maines MD. The heme oxygenase system: A regulator of second messenger gases. Annu Rev Pharmacol Toxicol 1997; 37:517-554.
23. Wang X, Hauptmann N, Taylor E et al. Neotrofin increases heme oxygenase-1 selectively in neurons. Brain Res 2003; 962:1-14.
24. Stocker R, Yamamoto Y, McDonagh AF et al. Bilirubin is an antioxidant of possible physiological importance. Science 1987; 235:1043-1046.
25. Mancuso C, Tringali G, Grossman A et al. The generation of nitric oxide and carbon monoxide produces opposite effects on the release of immunoreactive interleukin-1beta from the rat hypothalamus in vitro: Evidence for the involvement of different signaling pathways. Endocrinology 1998; 139:1031-1037.
26. Otterbein LE, Bach FH, Alam J et al. Carbon monoxide has anti-inflammatory effects involving the mitogen-activated protein kinase pathway. Nat Med 2000; 6:422-428.
27. Mancuso C, Bonsignore A, Di Stasio E et al. Bilirubin and s-nitrosothiols interaction: Evidence for a possible role of bilirubin as a scavenger of nitric oxide. Biochem Pharmacol 2003; 66:2355-2363.
28. Turner CP, Panter SS, Sharp FR. Anti-oxidants prevent focal rat brain injury as assessed by induction of heat shock proteins (HSP70, HO-1/HSP32, HSP47) following subarachnoid injections of lysed blood. Brain Res Mol Brain Res 1999; 65:87-102.
29. Motterlini R, Foresti R, Bassi R et al. Endothelial heme oxygenase-1 induction by hypoxia. Modulation by inducible nitric-oxide synthase and S-nitrosothiols. J Biol Chem 2000; 275:13613-13620.
30. Hill-Kapturczak N, Sikorski EM, Voakes C et al. An internal enhancer regulates heme and cadmium-mediated induction of human heme oxygenase-1. Am J Physiol Renal Physiol 2003; 285:F515-523.
31. Sun J, Hoshino O, Takaku K et al. Hemoprotein Bach1 regulates enhancer availability of heme oxygenase-1 gene. EMBO J 2002; 21:5216-5224.
32. Stewart D, Killeen E, Naquin R et al. Degradation of transcription factor Nrf2 via the ubiquitin-proteasome pathway and stabilization by cadmium. J Biol Chem 2003; 278:2396-2402.
33. Scapagnini G, D'Agata V, Calabrese V et al. Gene expression profiles of heme oxygenase isoforms in the rat brain. Brain Res 2002; 954:51-59.
34. Balogun E, Hoque M, Gong P et al. Curcumin activates the haem oxygenase-1 gene via regulation of Nrf2 and the antioxidant-responsive element. Biochem J 2003; 371:887-895.
35. Liu N, Wang X, McCoubrey WK et al. Developmentally regulated expression of two transcripts for heme oxygenase-2 with a first exon unique to rat testis: Control by corticosterone of the oxygenase protein expression. Gene 2000; 241:175-183.
36. Raju VS, McCoubrey WK, Maines MD. Regulation of heme oxygenase-2 by glucocorticoids in neonatal rat brain: Characterization of a functional glucocorticoid response element. Biochim Biophys Acta 1997; 1351:89-104.

37. Simonian NA, Coyle JT. Oxidative stress in neurodegenerative diseases. Annu Rev Pharmacol Toxicol 1996; 36:83-106.
38. Sayre LM, Smith MA, Perry G. Chemistry and biochemistry of oxidative stress in neurodegenerative disease. Curr Med Chem 2001; 8:721-738.
39. Andersen JK. Oxidative stress in neurodegeneration: Cause or consequence? Nat Med 2004; 10(Suppl):S18-25.
40. Mancuso C. Heme oxygenase and its products in the nervous system. Antioxid Redox Signal 2004; 6:878-887.
41. Panahian N, Yoshiura M, Maines MD. Overexpression of heme oxygenase-1 is neuroprotective in a model of permanent middle cerebral artery occlusion in transgenic mice. J Neurochem 1999; 72:1187-1203.
42. Takeda A, Perry G, Abraham NG et al. Overexpression of heme oxygenase in neuronal cells, the possible interaction with Tau. J Biol Chem 2000; 275:5395-5399.
43. Premkumar DR, Smith MA, Richey PL et al. Induction of heme oxygenase-1 mRNA and protein in neocortex and cerebral vessels in Alzheimer's disease. J Neurochem 1995; 65:1399-1402.
44. Schipper HM. Heme oxygenase-1: Role in brain aging and neurodegeneration. Exp Gerontol 2000; 35:821-830.
45. Calabrese V, Butterfield DA, Giuffrida Stella AM. Nutritional antioxidants and the heme oxygenase pathway of stress tolerance: Novel targets for neuroprotection in Alzheimer's disease. It J Biochemistry 2003b; 52:72-76.
46. Scapagnini G, Foresti R, Calabrese V et al. Caffeic acid phenethyl ester and curcumin: A novel class of heme oxygenase-1 inducers. Mol Pharmacol 2002; 61:554-561.
47. Scapagnini G, Butterfield DA, Colombrita C et al. Ethyl ferulate, a lipophilic polyphenol, induces HO-1 and protects rat neurons against oxidative stress. Antioxid Redox Signal 2004; 6:811-818.
48. Butterfield D, Castegna A, Pocernich C et al. Nutritional approaches to combat oxidative stress in Alzheimer's disease. J Nutr Biochem 2002; 13:444-461.
49. Motterlini R, Foresti R, Bassi R et al. Curcumin, an antioxidant and anti-inflammatory agent, induces heme oxygenase-1 and protects endothelial cells against oxidative stress. Free Radic Biol Med 2000b; 28:1303-1312.
50. Baranano DE, Rao M, Ferris CD et al. Biliverdin reductase: A major physiologic cytoprotectant. Proc Natl Acad Sci USA 2002; 99:16093-16098.
51. Kaur H, Hughes MN, Green CJ et al. Interaction of bilirubin and biliverdin with reactive nitrogen species. FEBS Lett 2003; 543:113-119.
52. Ganguli M, Chandra V, Kamboh MI et al. Apolipoprotein E polymorphism and Alzheimer disease: The Indo-US Cross- National Dementia Study. Arch Neurol 2000; 57:824-830.
53. Lim GP, Yang F, Chu T et al. Ibuprofen suppresses plaque pathology and inflammation in a mouse model for Alzheimer's disease. J Neurosci 2000; 20:5709-5714.
54. Kitamuro T, Takahashi K, Ogawa K et al. Bach1 functions as a hypoxia-inducible repressor for the heme oxygenase-1 gene in human cells. J Biol Chem 2003; 278:9125-9133.
55. Nakayama M, Takahashi K, Kitamuro T et al. Repression of heme oxygenase-1 by hypoxia in vascular endothelial cells. Biochem Biophys Res Commun 2000; 271:665-671.
56. Okinaga S, Takahashi K, Takeda K et al. Regulation of human heme oxygenase-1 gene expression under thermal stress. Blood 1996; 87:5074-5084.
57. Takahashi K, Nakayama M, Takeda K et al. Suppression of heme oxygenase-1 mRNA expression by interferon-gamma in human glioblastoma cells. J Neurochem 1999; 72:2356-2361.
58. Oyake T, Itoh K, Motohashi H et al. Bach proteins belong to a novel family of BTB-basic leucine zipper transcription factors that interact with MafK and regulate transcription through the NF-E2 site. Mol Cell Biol 1996; 16:6083-6095.
59. Sun J, Hoshino H, Takaku K et al. Hemoprotein Bach1 regulates enhancer availability of heme oxygenase-1 gene. EMBO J 2002; 21:5216-5224.
60. Shibahara S. The heme oxygenase dilemma in cellular homeostasis: New insights for the feedback regulation of heme catabolism. Tohoku J Exp Med 2003; 200:167-186.
61. Calabrese V, Boyd-Kimball D, Scapagnini G et al. Nitric oxide and cellular stress response in brain aging and neurodegenerative disorders: The role of vitagenes. In Vivo 2004c; 18:245-268.
62. Kostoglou-Athanassiou I, Forsling ML, Navarra P et al. Oxytocin release is inhibited by the generation of carbon monoxide from the rat hypothalamus—further evidence for carbon monoxide as a neuromodulator. Brain Res Mol Brain Res 1996; 42:301-306.
63. Mancuso C, Kostoglou-Athanassiou I, Forsling ML et al. Activation of heme oxygenase and consequent carbon monoxide formation inhibits the release of arginine vasopressin from rat hypothalamic explants. Molecular linkage between heme catabolism and neuroendocrine function. Brain Res Mol Brain Res 1997a; 50:267-276.

64. Mancuso C, Ragazzoni E, Tringali G et al. Inhibition of heme oxygenase in the central nervous system potentiates endotoxin-induced vasopressin release in the rat. J Neuroimmunol 1999; 99:189-194.
65. Parkes D, Kasckow J, Vale W. Carbon monoxide modulates secretion of corticotropin-releasing factor from rat hypothalamic cell cultures. Brain Res 1994; 646:315-318.
66. Pozzoli G, Mancuso C, Mirtella A et al. Carbon monoxide as a novel neuroendocrine modulator: Inhibition of stimulated corticotropin-releasing hormone release from acute rat hypothalamic explants. Endocrinology 1994; 135:2314-2317.
67. Mancuso C, Pistritto G, Tringali G et al. Evidence that carbon monoxide stimulates prostaglandin endoperoxide synthase activity in rat hypothalamic explants and in primary cultures of rat hypothalamic astrocytes. Brain Res Mol Brain Res 1997; 45:294-300.
68. Jaggar JH, Leffler CW, Cheranov SY et al. Carbon monoxide dilates cerebral arterioles by enhancing the coupling of Ca^{2+} sparks to Ca^{2+}-activated K^+ channels. Circ Res 2002; 91:610-617.
69. Ryter SW, Otterbein LE, Morse D et al. Heme oxygenase/carbon monoxide signaling pathways: Regulation and functional significance. Mol Cell Biochem 2002; 234-235:249-263.
70. Yenari MA, Giffard RG, Sapolsky RM et al. The neuroprotective potential of heat shock protein 70 (HSP70). Mol Med Today 1999; 5:525-531.
71. Yenari MA. Heat shock proteins and neuroprotection. Adv Exp Med Biol 2002; 513:281-299.
72. Hata R, Maeda K, Hermann D et al. Dynamics of regional brain metabolism and gene expression after middle cerebral artery occlusion in mice. J Cereb Blood Flow Metab 2000; 20:306-315.
73. Perez N, Sugar J, Charya S et al. Increased synthesis and accumulation of heat shock 70 proteins in Alzheimer's disease. Brain Res Mol Brain Res 1991; 1:249-254.
74. Yoo BC, Seidl R, Cairns N et al. Heat-shock protein 70 levels in brain of patients with Down syndrome and Alzheimer's disease. J Neural Transm Suppl 1999; 57:315-322.
75. Morrison-Bogorad M, Zimmerman AL, Pardue S. Heat-shock 70 messenger RNA levels in human brain: Correlation with agonal fever. J Neurochem 1995; 64:235-246.
76. Kakimura J, Kitamura Y, Takata K et al. Microglial activation and amyloid-beta clearance induced by exogenous heat-shock proteins. FASEB J 2002; 16:601-603.
77. Calabrese V, Copani A, Testa D et al. Nitric oxide synthase induction in astroglial cell cultures: Effect on heat shock protein 70 synthesis and oxidant/antioxidant balance. J Neurosci Res 2000a; 60:613-622.
78. Calabrese V, Testa G, Ravagna A et al. HSP70 induction in the brain following ethanol administration in the rat: Regulation by glutathione redox state. Biochem Biophys Res Commun 2000b; 269:397-400.
79. Calabrese V, Bates TE, Giuffrida Stella AM. NO synthase and NO-dependent signal pathways in brain aging and neurodegenerative disorders: The role of oxidant/antioxidant balance. Neurochem Res 2000c; 65:1315-1341.
80. Calabrese V, Scapagnini G, Giuffrida Stella AM et al. Mitochondrial involvement in brain function and dysfunction: Relevance to aging, neurodegenerative disorders and longevity. Neurochem Res 2001; 26:739-764.
81. Iverson SL, Orrenius S. The cardiolipin-cytochrome c interaction and the mitochondrial regulation of apoptosis. Arch Biochem Biophys 2004; 423:37-46.
82. Calabrese V, Ravagna A, Colombrita C et al. Acetylcarnitine induces heme oxygenase in rat astrocytes and protects against oxidative stress: Involvement of the transcription factor Nrf2. J Neurosci Res 2005; 79:509-521.
83. Rai G, Wright G, Scott L et al. Double-blind, placebo controlled study of acetyl-L-carnitine in patients with Alzheimer's dementia. Curr Med Res Opin 1990; 11:638-641.
84. Patrick L. Nutrients and HIV: Part three - N-acetylcysteine, alpha-lipoic acid, L-glutamine, and L-carnitine. Altern Med Rev 2000; 5:290-305.
85. Calabrese V, Scapagnini G, Ravagna A et al. Disruption of thiol homeostasis and nitrosative stress in the cerebrospinal fluid of patients with active multiple sclerosis: Evidence for a protective role of acetylcarnitine. Neurochem Res 2003c; 28:1321-1328.
86. Albertini P, Amenta F, Bossoni G et al. Effect of acetyl-L-carnitine treatment on the density of muscarinic receptors in the brain of methylazoxymethanol-microencephalic rats. Drugs Exp Clin Res 1989; 15:421-7.
87. Rao KV, Mawal YR, Qureshi IA. Progressive decrease of cerebral cytochrome C oxidase activity in sparse-fur mice: Role of acetyl-L-carnitine in restoring the ammonia-induced cerebral energy depletion. Neurosci Lett 1997; 224:83-86.
88. Carta A, Calvani M, Bravi D et al. Acetyl-L-carnitine and Alzheimer's disease: Pharmacological considerations beyond the cholinergic sphere. Ann NY Acad Sci 1993; 695:324-326.

89. Ando S, Kobayashi S, Waki H et al. Animal model of dementia induced by entorhinal synaptic damage and partial restoration of cognitive deficits by BDNF and carnitine. J Neurosci Res 2002; 70:519-527.
90. Chiechio S, Caricasole A, Barletta E et al. L-Acetylcarnitine induces analgesia by selectively up-regulating mGlu2 metabotropic glutamate receptors. Mol Pharmacol 2002; 61:989-996.
91. Spagnoli A, Lucca U, Menasce G et al. Long-term acetyl-L-carnitine treatment in Alzheimer's disease. Neurology 1991; 41:1726.
92. Caprioli A, Markowska AL, Olton DS. Acetyl-L-Carnitine: Chronic treatment improves spatial acquisition in a new environment in aged rats. J Gerontol 1995; 50:232-236.
93. Chauhan NB, Siegel GJ. Effect of PPF and ALCAR on the induction of NGF- and p75-mRNA and on APP processing in Tg2576 brain. Neurochem Int 2003; 43:225-233.
94. Calabrese V, Scapagnini G, Latteri S et al. Long-term ethanol administration enhances age-dependent modulation of redox state in different brain regions in the rat: Protection by acetyl carnitine. Int J Tissue React 2002d; 24:97-104.
95. Hagen TM, Ingersoll RT, Wehr CM et al. Acetyl-L-carnitine fed to old rats partially restores mitochondrial function and ambulatory activity. Proc Natl Acad Sci USA 1998; 95:9562-9566.
96. Wolff RL, Combe NA, Entressangles B. Positional distribution of fatty acids in cardiolipin of mitochondria from 21-day-old rats. Lipids 1985; 20:908-914.
97. Armeni T, Principato G, Quiles JL et al. Mitochondrial dysfunctions during aging: Vitamin E deficiency or caloric restriction—two different ways of modulating stress. J Bioenerg Biomembr 2003; 35:181-191.
98. Poon HF, Shepherd HM, Reed T et al. Proteomics analysis provides insight into caloric restriction-mediated reduced oxidation and altered expression of brain proteins associated with age-related impaired cellular processes: Mitochondrial dysfunction, glutamate dysregulation and impaired protein systhesis. Neurobiol Aging 2005; in press.
99. Wheeler DS, Dunsmore KE, Wong HR. Intracellular delivery of HSP70 using HIV-1 Tat protein transduction domain. Biochem Biophys Res Commun 2003; 301:54-59.
100. Pocernich CB, Sultana R, Hone E et al. Effects of apolipoprotein E on the human immunodeficiency virus protein tat in neuronal cultures and synaptosomes. J Neurosci Res 2004; 77:532-539.
101. Weng YH, Yang G, Weis S et al. Interaction between heme oxygenase-1 and 2 proteins. J Biol Chem 2003; 278:50999-51005.
102. Sultana R, Ravagna A, Mohmmad-Abdul H et al. Ferulic acid ethyl ester protects neurons against amyloid beta-peptide(1-42)-induced oxidative stress and neurotoxicity: Relationship to antioxidant activity. J Neurochem 2005; 92:749-758.
103. Otterbein LE, Zuckerbraun BS, Haga M et al. Carbon monoxide suppresses arteriosclerotic lesions associated with chronic graft rejection and with balloon injury. Nat Med 2003; 9:183-190.

Chapter 5

Heat Shock Proteins and the Regulation of Apoptosis

Una FitzGerald, Adrienne M. Gorman and Afshin Samali*

Abstract

Since the elucidation of their functions in protein folding and translocation, heat shock protein chaperones have been a target of research in all spheres of biomedicine. Within the last five years, research efforts have intensified, following the discovery of raised levels of heat shock protein (Hsp) expression in the brains of patients suffering from many neurodegenerative disorders, including Alzheimer's, Parkinson's and Huntington's diseases and cerebral ischaemia. Expression of Hsps in the brains of patients is thought to form part of a general protective stress response. The stress in question, however, varies, depending on the particular disease. For example, accumulation of α-synuclein aggregates in Parkinson's Disease causes stress to the protein folding machinery of the cells, with consequent up-regulation of stress proteins including Hsps. When markers indicative of the occurrence of apoptosis were also found in degenerating brain tissue, the question of how heat shock proteins might impact on apoptotic neural cells was raised. However, their particular function under diseased conditions remains unclear. This chapter highlights the involvement of Hsps in the regulation of neural apoptosis, from the original reports of Hsp expression during neurological disorders, to evidence of their neuroprotective properties and their potential as therapeutic molecules.

Introduction to Apoptosis

Within tissues cell death can occur via either necrosis or apoptosis. Necrosis is a passive form of cell death occurring mainly under pathological conditions, where a rapid loss of ion-flux control leads to the swelling and rupture of the cell and its organelles. In contrast, apoptosis is a controlled, energy-dependent form of cell death involving an elaborate network of signal transduction pathways in both its initiation and execution. Characteristic hallmarks of apoptosis include membrane blebbing, condensation of nuclear chromatin, cytoplasmic shrinkage, nuclear fragmentation and formation of apoptotic bodies. Apoptosis is induced in response to a large variety of stimuli including cytokines, cytotoxic drugs, oxidative stress and ionising radiation.[1] These diverse stimuli trigger apoptosis by activating one or more signal transduction pathways, which then converge to activate a conserved family of aspartic-acid specific cysteine proteases, referred to as caspases.[2] Caspases are constitutively expressed within cells as inactive precursor zymogens and are activated in response to apoptotic stimuli by changes in the three dimensional structure of the protein or by specific proteolytic cleavage.[3-5] Once activated they orchestrate the demise of the cell through the cleavage of a specific subset of cellular substrates,[6] resulting in the characteristic biochemical and morphological changes associated with apoptosis.[7]

*Corresponding Author: Afshin Samali—NCBES, National University of Ireland, Galway, University Road, Galway, Ireland, Email: afshin.samali@nuigalway.ie

Heat Shock Proteins in Neural Cells, edited by Christiane Richter-Landsberg.

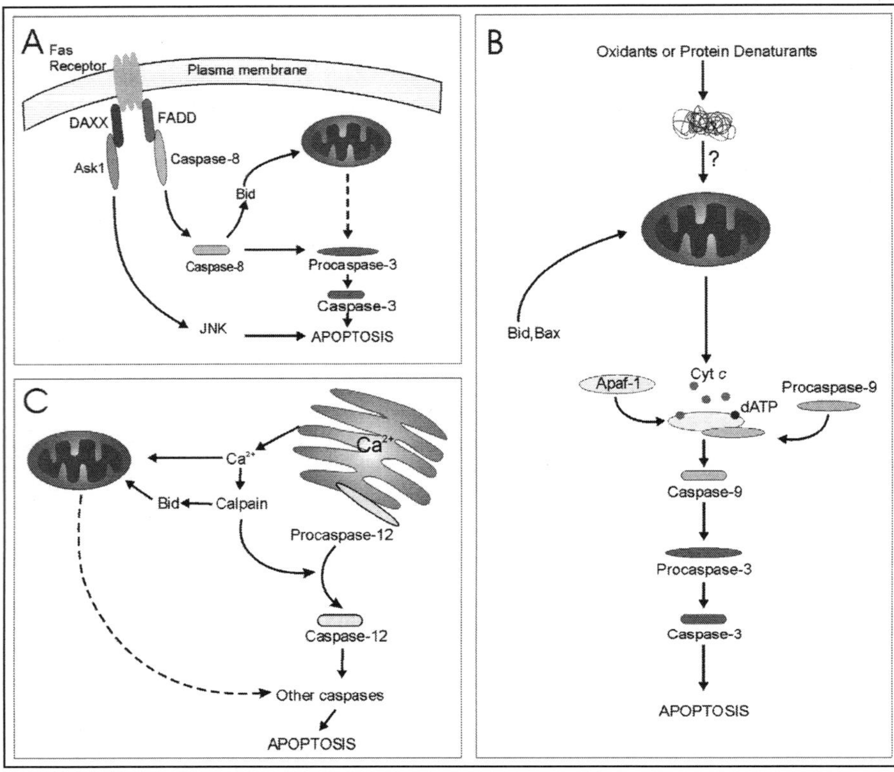

Figure 1. Caspases can be activated by at least three different pathways leading to apoptosis. A) Mitochondrial pathway is initiated due to release of cytochrome *c* from the mitochondrial intermembrane space into the cytosol, leading to the formation of the apoptososme which culminates in the activation of caspase-9. B) The death receptor pathway, activated by trimerization of death receptors at the plasma membrane (e.g., Fas receptor), which leads to autoactivation of caspase-8. Daxx, a nuclear protein, translocates to the membrane during Fas-mediated apoptosis. Daxx binds at one end to the Fas receptor and at the other with Ask1, thus mediating a caspase-independent cell death. There is a cross-talk between the death receptor pathway and the mitochondria, through cleavage of the protein Bid, by caspase-8. C) The ER stress pathway is activated when unfolded proteins accumulate in the ER or when there is disruption in the ER Ca^{2+} homeostasis, which leads to activation of caspase-12, possibly by a Ca^{2+}-mediated process involving calpain. Ca^{2+} may also amplify the pathway by acting on mitochondria.

Caspase activation during apoptosis can proceed via a number of different pathways (Fig. 1). The extrinsic pathway involves the binding of death ligands to cell surface receptors (e.g., Fas/CD95/Apo-1 or TNF receptor) resulting in the recruitment of the adaptor molecules Fas Associated Death Domain (FADD) or TNF Receptor Associated Death Domain (TRADD) to the cytosolic end of the receptor leading to the formation of the Death Inducing Signalling Complex (DISC) at the plasma membrane and resultant activation of pro-caspase-8 and thereby pro-caspase-3.[8-10] In addition to FADD, the DAXX adaptor molecule may also transduce death signals to JNK, via apoptosis signal-regulating kinase 1 (ASK1) recruitment to the DISC. As JNK is a substrate of ASK1, this broadens the effect of Fas ligand binding to include JNK-mediated apoptosis (Fig. 1A).[11]

The intrinsic pathway is initiated through the release of cytochrome *c* from the intermembrane space of mitochondria (Fig. 1B).[12,13] Cytochrome *c* translocates across the outer mitochondrial membrane, by a number of possible pathways[14] to the cytosol, where it binds to Apoptosis Protease Activating Factor 1 (Apaf-1) and in the presence of dATP (or ATP) cytochrome *c*/dATP facilitates Apaf-1 oligomerization and the recruitment of pro-caspase-9.[15] The formation of this caspase-activating complex, termed the apoptosome, results in the activation of pro-caspase-9, which further amplifies the caspase cascade by its ability to process effector caspases including pro-caspase-3.[16]

The endoplasmic reticulum (ER) stress-induced pathway is a more recent and controversial apoptotic program, originally identified when ER-associated caspase-12 was cleaved in response to chemical inducers of ER stress.[17-20] Although the origin of the particular ER disturbance may differ, it activates a common signalling pathway termed the unfolded protein response (UPR).[21,22] Three ER trans-membrane proteins initiate the UPR and these are: an ER-associated type I trans-membrane protein kinase (Ire1), activated transcription factor 6 (ATF6) and PKR-like endoplasmic reticulum kinase (PERK). In resting cells, the 78 kDa glucose-regulated protein (Grp78 or Bip) binds the ER-luminal ends of these molecules but in the presence of ER stress, it detaches from Ire1, PERK and ATF6, thus, enabling their activation. PERK activation leads to a temporary halt in protein translation and under certain conditions, the activation of the pro-apoptotic protein Gadd153 or CHOP. Active Ire1 causes induction of the X-box binding 1 (XBP1) transcription factor, thought to induce expression of itself and a subset of other proteins involved in the UPR.[23] ATF6, activated through cleavage, activates transcription of chaperone proteins and transcription factors, e.g., Grp78 and XBP1. Controversy surrounding ER stress-induced apoptosis has centred on the lack of expression of caspase-12 in human cells.[24,25] However, the search for the functional homologues of human caspase-12 has indicated a possible role for caspase-4 in mediating apoptosis triggered by the ER (Fig. 1C).[26]

The mitochondrion has been identified as a central control point for the integration of diverse death signals during apoptosis. Many of the key molecules involved in apoptosis are located within or associated with the mitochondrion.[13] These include cytochrome *c*,[27] AIF,[28] Smac/Diablo,[29] EndoG,[30] Puma[31] and pro-caspase-3.[2,32] When released into the cytosol, these molecules function both in the initiation and execution of the apoptotic program. It is therefore unsurprising to find that the mitochondrion serves as a control point for cross-talk between the intrinsic, extrinsic and ER stress pathways of apoptosis. In this respect it is well established that the activation of pro-caspase-8 by death receptors can result in the cleavage of an endogenous cellular protein, Bid, generating a truncated pro-apoptotic fragment that translocates to the mitochondrion where it induces cytochrome *c* release.[33,34] This mitochondrial step is found to be critical in certain cell types where the engagement of death receptors alone is not sufficient to induce caspase activation without cytochrome *c* release.[35] Further reports also suggest that stimuli that induce DNA damage and subsequently apoptosis do so via translocation from the nucleus to the mitochondrion of molecules that stimulate the release of cytochrome *c*.[36,37] Naturally, since the mitochondrion has such a key role during apoptosis many of the endogenous cellular proteins that function as crucial inhibitors of cell death execute their anti-apoptotic capabilities by acting on mitochondria and preventing the release of crucial pro-apoptotic proteins.

To date the best-characterised endogenous protein modulators of apoptosis are the members of the Bcl-2 family of proteins.[38] Both pro- and anti-apoptotic members of this family have been identified and studied within mammalian cells, where they are seen to function as key determinants of cell fate. However, the recent correlation between the expression of Hsps and increased cell survival, has pointed to Hsps as playing a critical role in the regulation of the apoptotic machinery.

Hsp27 and Hsp70, the Anti-Apoptotic Heat Shock Proteins

Members of the heat shock family of proteins can be divided into two camps. The first contains chaperones, that are constitutively expressed and whose normal function is aiding protein folding and maturation (e.g., Hsc70), but whose expression does not seem to readily

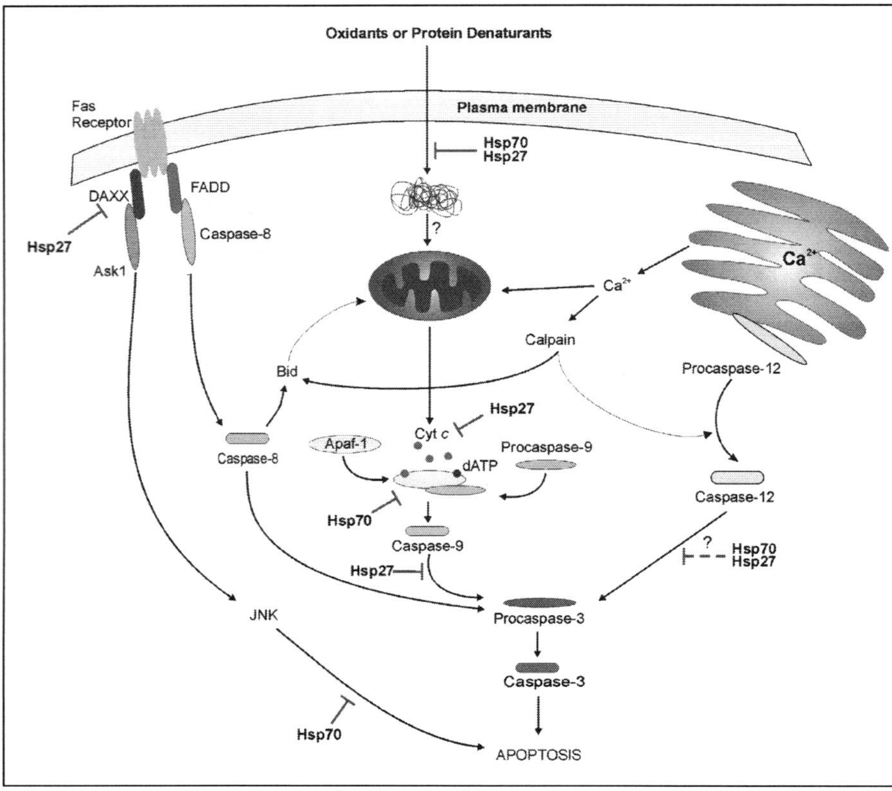

Figure 2. Heat shock protein regulation of apoptotic pathways. A mechanism by which apoptosis is initiated is due to changes in the intracellular redox balance and production of reactive oxygen species. This results in changes in the mitochondria and release of pro-apoptotic factors. Hsp27 and Hsp70 can maintain both redox homeostasis and mitochondrial stability in the cell. Also Hsp27 can bind to cytochrome *c*, after its release from mitochondria, and pro-caspase-3, thus, preventing apoptosome formation and events downstream of mitochondrial damage. In cell-free systems, Hsp70 has also been demonstrated to inhibit apoptosome formation. Apart from the caspase-dependent apoptosis, Hsp27 is reported to block Daxx-mediated cell death. Hsp27 can prevent the translocation of Daxx to membrane and its interaction with Fas. Currently there is no information regarding the role of Hsp27 or Hsp70 in the regulation of ER stress-mediated apoptosis.

change in response to stress. The second consists of Hsps which may, albeit at low levels, or may not be constitutively expressed in cells, but whose expression is significantly induced in response to stress. These latter proteins have been shown to have cytoprotective properties and exert their effect by inhibiting apoptosis. Hsp27 and Hsp70 are the main chaperones associated with an anti-apoptotic effect in vivo. The mechanisms by which they are thought to impinge on apoptosis, are described below and summarised in Figure 2.

Hsp70 and the Regulation of Apoptosis

There are many ways in which Hsp70 can inhibit the execution of caspase-dependent apoptosis and caspase-independent cell death. Most are centred on mitochondria-mediated events. For example, direct binding of the Apaf-1 CARD domain by Hsp70 was shown to prevent apoptosome formation and caspase-9 activation in cell-free systems.[39,40] However, recent findings suggest that inhibition of caspase activation by Hsp70 is due to interference with mitochondrial release of cytochrome *c*.[41] Through direct binding of another mitochondria-associated protein, AIF, Hsp70 can broaden its protective role by inhibiting caspase-independent nuclear fragmentation.[42,43] Although Hsp70 has been shown to be protective in a TNFα-induced cell death model, recent studies indicate that whether or not Hsp70 is protective in a given pathway, depends on whether or not activation of the mitochondrial pathway is required.[44] Hsp70 may also employ indirect mechanisms to inhibit amplification of death inducing signals. For example, Hsp70, by inhibiting JNK activity, prevents or down-regulates Bid-, p53- and/or c-myc-mediated release of cytochrome *c* and Smac/Diablo in response to stress.[45-48] However, this notion was contradicted in a different model system by Jaattela and colleagues who showed that cells over-expressing Hsp70 did not prevent activation of stress kinases including JNK, but rather inhibited the downstream effects of caspase activity.[49] Alternatively, Hsp70 may interfere with post-mitochondrial signalling, through direct binding of pro-caspases -3 and -7.[50]

In addition to the stress-activated JNK, Hsp70 has also been shown to directly bind ASK1, preventing its homo-oligomerization, with consequent protection of cells from oxidative stress and death through inhibition of ASK1-mediated cytochrome *c* release.[51]

The range of protective effects associated with Hsp70 may be altered, depending on whether or not Hsp70 is in complex with cochaperones including Hsp40 or BAG-1. Under normal conditions, Hsp70 exists as a monomer, executing house-keeping chaperone functions. However, under stressed conditions, it may be found in complex with Hsp40, thus, broadening its substrate specificity.[52,53] This is especially important when such a complex can alter the expected anti-apoptotic effect, to a pro-apoptotic one. For example, in complex with Hsp40, Hsp70 had been shown to enhance caspase-activated DNase, thus augmenting T-cell receptor-mediated apoptosis.[54] However, in line with general observations for Hsp70, while in complex with other cochaperones, e.g., DNaJ, Hsp70 protects cells by inhibiting Bax translocation to the mitochondria.[55]

Hsp27 and the Regulation of Apoptosis

In many ways, the protective effects of Hsp27 supplement those described for Hsp70. For example, while Hsp70 may prevent release of cytochrome *c* from mitochondria, phosphorylated Hsp27, through direct cytochrome *c* binding, inhibits apoptosome formation and subsequent activation of caspase-9.[56-59] In addition, Hsp27 has been shown to inhibit caspase-3 activity, by interacting with pro-caspase-3, preventing its cleavage by caspase-9.[57,60] At the level of mitochondria, Hsp27 may also exert its protective effect by inhibiting the release of SMAC/Diablo.[61]

Unlike Hsp70, Hsp27 has the potential to interfere directly with death-receptor death-inducing signals, without the need for the involvement of mitochondrial factors. This effect is achieved by phosphorylated dimers of Hsp27 which are capable of interacting with and inhibiting DAXX, which links the Fas receptor to Ask 1 and the downstream JNK pathway.[62] Another kinase linked to Hsp27's protective role, is Akt, a pro-survival molecule, which targets and inactivates Ask1.[63] Hsp27 interacts with Akt directly and has been proposed to enhance its activation through its chaperone function.[64,65] Hsp27 is itself a target of active Akt and once phosphorylated, it is released from the Akt complex and so becomes available to participate in other anti-apoptotic functions. A similar indirect protective activity has been proposed for Hsp27, through participation in a complex between ubiquitin and IkBα. IkBα functions to sequester the pro-survival transcription factor NF-κB. Phosphorylation of IkBα leads to its ubiquitination and subsequent degradation by the 26S proteasome. By direct binding of the proteasome, the ubiquitin chains and phosphorylated IκBα, Hsp27 facilitates the degradation of IkBα.[66] This releases NF-κB, enabling its translocation to the nucleus where it transactivates the expression of pro-survival genes.

Studies of Hsp27 function under conditions of oxidative stress-induced cell death, have indicated that Hsp27 achieves its protective effect by increasing glutathione levels,[67-69] reducing cytosolic reactive oxygen species (ROS),[68,70] raising glucose-6-phosphate dehydrogenase activity[71] and/or lowering cytosolic iron.[69]

The potential for Hsp27 to protect cells through preservation of the architecture of the cytoplasm has been the subject of much recent discussion. Based on studies of post-translational modification of Hsp27 in heat-shocked or stressed cells, stabilisation of actin filaments by the phosphorylated nonoligomeric form of Hsp27 has been implicated in its protective function.[59,72-74] Hsp27 has a known role in actin reorganisation, a phenomenon which is required for the formation of membrane blebs and apoptotic bodies that are generated during the course of apoptosis.[75,76] This role appears to contradict the suggested actin-protecting functions described above, raising the possibility that the particular actin-associated function of Hsp27 during apoptosis may be signal- and cell type-dependent. However, given that many structural proteins are cleaved by caspase-3 during apoptosis,[6] it is conceivable that Hsp27's ability to inhibit caspase-3 may be sufficient to maintain the cytoskeleton intact. In neurons, the cytoskeletal protective function of Hsp27 has been linked to its ability to decrease levels of phosphorylated tau and enhance nonubiquitin-dependent degradation of hyperphosphorylated tau.[77,78] Further details of the effects of Hsp27 on tau are provided in Chapters 2 and 6.

Apoptosis, Hsps and Neurodegenerative Disorders

That apoptosis occurs during the course of the neurodegenerative disease process, is well documented. Caution is, however, required when interpreting data derived from human tissue, as artefactual occurrence of DNA degradation during tissue processing may lead to false conclusions. In most cases, detection of nuclear fragmentation or cleaved forms of caspases or their substrates is used to detect apoptosis. For example, in neurons and glia, terminal transferase-mediated biotinylated-UTP nick end-labelling (TUNEL) detected raised level of fragmented DNA in brain tissue derived from Alzheimer's[79-82] and Huntington's Disease patients.[81,83] More recently, TUNEL positivity was detected in addition to raised levels of Bax, Bcl-2, cleaved caspase-3 and cleaved poly(ADP-ribose) polymerase (PARP) in Huntington's Disease brain tissue.[84] A very useful discussion of some of the issues surrounding the use of DNA fragmentation as an apoptosis marker is presented by Tatton et al (2003),[85] who also confirmed the occurrence of DNA fragmentation, Bax up-regulation and caspase-3 cleavage during Parkinson's Disease (PD).[86] Additional studies of PD tissue have shown the raised levels of apoptosis signalling molecules including the Fas receptor[87] and caspases -3, -8 and -9.[88,89]

Neurodegenerative diseases, whether acute or chronic, are a major medical problem in the aging population. Therapies are rare and applied usually during the late stages of the disease, when a vast number of cells are already lost. Many efforts have been made to develop new strategies to treat these disorders, but so far, there has been no breakthrough. A characteristic shared by some experimental neuroprotective substances is the induction of the Hsps and in particular Hsp70 and Hsp27. In response to many metabolic disturbances and injuries cells mount a stress response with induction of a variety of proteins, most notably inducible Hsps. There is evidence that upon injury to the brain Hsp70 is induced to high levels in brain regions that are relatively resistant to injury. With the availability of transgenic animals and gene transfer, it has become increasingly clear that selective over-expression of Hsps leads to protection in several different models of nervous system injury that involve neurodegeneration and apoptosis, including ischemia/excitotoxicity and models of Parkinson's and Alzheimer's diseases. Some of these will be discussed.

Ischemia/Excitotoxicity

About 700,000 Americans suffer ischemic stroke each year. That is, one every 45 seconds (American Stroke Society statistics). The brain is particularly vulnerable to ischemic damage and even transient interruption to blood flow can cause significant neuronal cell death. Considerable evidence supports a role for apoptosis in cerebral ischemia. While damaged neurons often die from necrosis, apoptosis contributes significantly to cell death subsequent to cerebral ischemia, with apoptosis being predominant when the insult is relatively mild. Hippocampal neurons are particularly susceptible to cell damage and this is largely mediated through over-excitation of glutamate receptors as a result of uncontrolled glutamate release. This latter phenomenon is termed excitotoxicity. Oxygen-glucose deprivation and exposure of neurons to glutamate are both used to mimic neuronal cell death during ischemia.

Overexpression of Hsp70 has been shown to provide protection from cerebral ischaemia both in animal models of stroke and in cell culture models and has been shown to involve anti-apoptotic, anti-necrotic and anti-protein aggregation mechanisms.[90] For example, comparison of the effect of an ischemic insult on Hsp70 transgenic mice versus their wild-type litter-mates demonstrated significant neuroprotection in transgenic animals.[91] Similarly, Hsp70 over-expression protects hippocampal neurons from global cerebral ischemia, and this protection may be mediated in part by increased Bcl-2 expression.[92] In another example of Hsp70-mediated protection it was shown that primary cultures of dorsal root ganglia (DRG) sensory neurons can be protected against subsequent severe thermal or ischaemic stress by mild thermal or ischaemic preconditioning or by over-expression of Hsp70.[93] In an excitotoxic model of retinal ischaemia, intravitreal injection of NMDA, a glutamate receptor agonist, was used to induce apoptosis in retinal ganglion cells. Again, hyperthermic preconditioning leading to elevated Hsp70 reduced the number of TUNEL-positive cells in the RGCL.[94]

Astrocytes perform many functions that protect neurons during stress, including transmitter uptake, metabolic support, and protection from oxidative stress. However, their protective role can be enhanced by over-expressing Hsp70 in astrocytes. This was elegantly demonstrated by Gifford and colleagues, when murine cortical astrocytes transfected with Hsp70 were cocultured with neurons and shown to protect them from the effects of combined oxygen-glucose deprivation, or glucose deprivation.[95]

Protection from death caused by ischaemia can also be achieved through over-expression of Hsp27, as demonstrated when rat retinal ganglion cells were transfected with Hsp27.[96]

Parkinson's Disease

Mitochondrial dysfunction and oxidative stress have been implicated in PD. In addition, genetic evidence points to an important role for protein misfolding, aggregation, and failure in the proteasomal degradation of specific neuronal proteins in the pathogenesis of PD. Recently, Hsp70 gene transfer to dopaminergic neurons by a recombinant adeno-associated virus (AAV) was found to significantly protect the mouse dopaminergic system against MPTP-induced neuron loss and the associated decline in striatal dopamine levels and tyrosine hydroxylase-positive fibers.[97]

A variety of in vitro model systems have also been used to investigate molecular mechanisms of PD, including cell lines such as PC12, MN9D and primary cultures of mesencephalic neurons. For example, over-expression of Hsp70 protected MES (mesencephalic/neuroblastoma) cells from rotenone-mediated cytotoxicity and decreased soluble α-synuclein aggregation.[98] In this study, the protection afforded by Hsp70 transfection was related to suppression of rotenone-induced oxidative stress as well as mitochondrial and proteasomal dysfunction. Heat shock, leading to the induction of both Hsp70 and Hsp25, has also been shown to protect PC12 cells against cell death by the Parkinson mimetics, MPP^+ and 6-hydroxydopamine (Fig. 3).[99,100] Moreover, the induction of Hsp25 in PC12 cells following exposure to 6-OHDA was associated with cell survival and over-expression of human Hsp27 in these cells attenuated 6-OHDA-induced apoptosis.[100]

Intracellular proteinaceous inclusions called Lewy bodies (LB) are the histological hallmarks of PD, and are primarily composed of misfolded aggregates of α-synuclein into prefibrillar and fibrillar species. Aggregation and cytotoxicity of misfolded α-synuclein is postulated to be crucial in the disease process of PD and DLB (dementia with Lewy bodies). Hsp70 has been shown to inhibit α-synuclein fibril formation via preferential binding to prefibrillar species and to alter the characteristics of toxic α-synuclein full aggregates.[101]

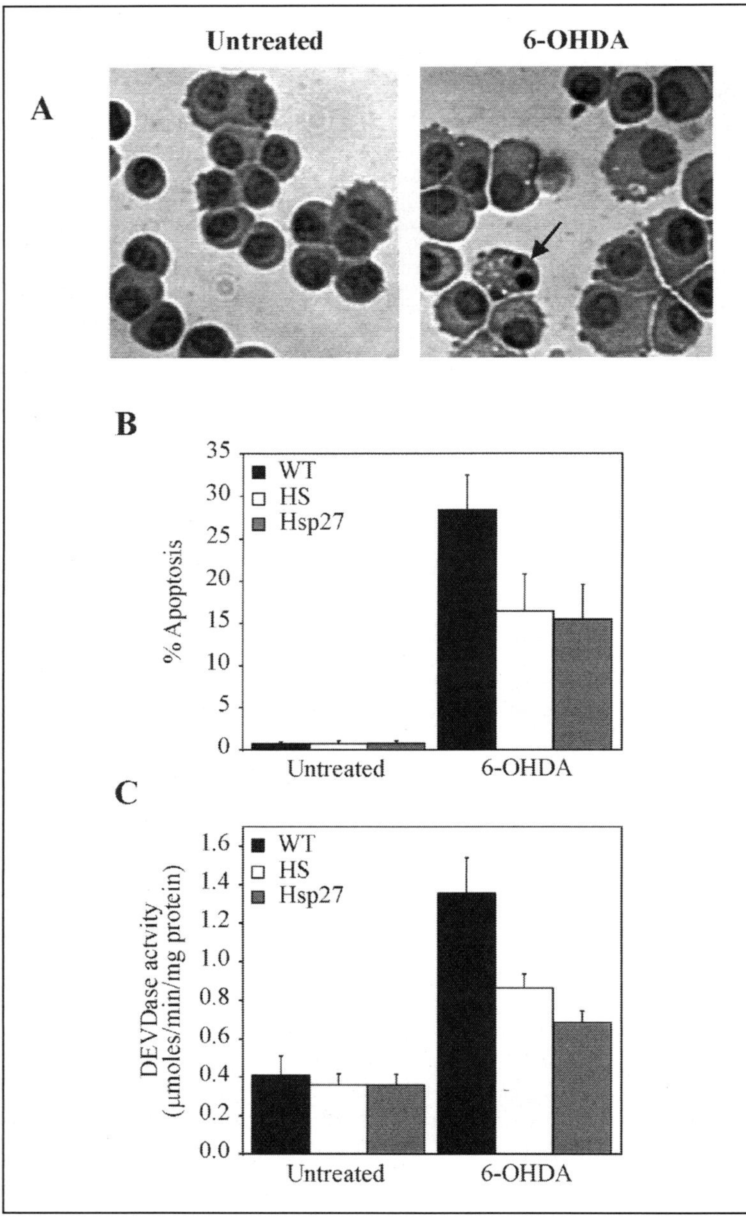

Figure 3. Inhibition of 6-hydroxydopamine (6-OHDA)-induced apoptosis by heat shock and overexpression of Hsp27. A. PC12 cells were exposed to 200 µM 6-OHDA for 24 h. Cytocentrifuge preparations of the cells were made and stained with haematoxylin and eosin. The arrow indicates a cell with a condensed and fragmented nucleus characteristic of apoptosis. B and C. In order to investigate the effect of Hsps on the induction of apoptosis, cells were either untreated (WT), heat shocked at to 41.5 °C for 1 h (HS) or were transfected with Hsp27 (Hsp27) prior to treatment with 200 µM 6-OHDA for 24 h. Apoptosis was determined by staining cells and counting those with condensed nuclei (B) and by assessing DEVDase caspase activity in whole cell extracts (C).

Table 1. Neuroprotective strategies exploiting Hsp70

Model System	Disorder Targeted	Mode of Overexpression	Comment	Reference
HT22 mouse hippocampal neurons	General neurodegen-eration	Retroviral transduction	Partial protection from glutamate toxicity achieved.	106
Primary astrocytes	Ischaemia	HSV-mediated infection	Geldanamycin treatment duplicated effect of infection of cells	90
Primary astrocytes and neurons from Hsp70 transgenic mouse	Ischaemia	Transgene expression	Primary astrocytes and hippocampal but not cortical neurons were protected from ischaemia. Whole animal not protected.	107
Dorsal root ganglion neurons	Ischaemia	Thermal Pre-conditioning and transfection	Survival of DRG cells following lethal thermal shock or ischaemia was increased by Hsp70 overexpression.	93
Transgenic mouse	Ischaemia	Transgene overexpression	Reduced mortality and brain damage in Hsp70 transgenic mice. Suggested role for HSP interaction with AIF.	91
Animal subjected to ischaemia	Ischaemia	HSV-mediated infection	Hippocampal CA1 neurons protected in vivo from affects of ischaemia.	90
Transgenic mouse	Ischaemia	Transgene expression	Infarction volume reduced following MCAO blockade.	108
Transgenic mouse	Ischaemia	Transgene expression	Hippocampal neurons showed no pyknosis after MCAO occlusion.	109
Intravitreal injection of NMDA	Ischaemia and excito-toxicity	Thermal Pre-conditioning	Reduced number of TUNEL-positive retinal ganglion cells.	94
MPTP mouse	Parkinson's disease	Adenoviral infection	Neuronal loss in substantia nigra reduced.	97
Transgenic mouse and human H4 neuroglioma cells	Parkinson's disease	Transgene expression and cellular transfection	Reduced α-synuclein aggregation and cytotoxicity in the presence of Hsp70 overexpression.	102
Mixed rat primary cortical cultures	Alzheimer's disease	Viral infection	Hsp70 over-expression rescued neurons in vitro from the toxic effects of β-amyloid accumulation.	104

Table 2. Neuroprotective strategies exploiting Hsp27

Model System	Disorder Targeted	Mode of Over-Expression	Comment	Reference
RGC-5 rat retinal ganglion cell line	Ischaemia	Transfection	Cells protected from serum, oxygen and glucose reduction and calcium overload.	96
HCN2A neuronal cells	Alzheimer's disease	Chemical delivery	Hsp27 reduced levels of hyperphosphorylated tau and enhanced its degradation.	77
ND7 DRG dorsal root ganglion/ neuroblastoma fusion cells	Parkinson's disease	HSV-mediated infection	Reduced caspase activity detected. Hsp70 was not protective against same insults, with exception of ischaemia.	103
PC12 rat pheochrom-ocytoma cells	Parkinson's disease	Transfection	Increased survival of 6-OHDA-treated cells over-expressing Hsp27.	100

In an in vivo model, breeding α-synuclein transgenic mice with Hsp70-overexpressing mice led to a significant reduction in α-synuclein aggregation and toxicity of both the high molecular weight and detergent-insoluble α-synuclein species.[102]

Similarly, Hsp27 has been shown to have a potent anti-apoptotic effect against the damage caused by wild-type and mutant α-synuclein in mammalian neuronal cells.[103] However, it is intriguing to note that the same study did not find a similar protection by Hsp70.

Alzheimer's Disease

There is comparatively little evidence to indicate that Hsps may have protective effects in Alzheimer's disease. Expression of amyloid beta peptide has been reported to induce Hsp70 in neurons and in the same study Hsp70 overexpression rescued neurons from the toxic effects of intracellular amyloid beta accumulation.[104] Estrogen and androgens have also been shown to protect against intracellular amyloid beta toxicity through inducing Hsp70.[105] Thus, Hsp70 may have potential in neuroprotection in models of Alzheimer's disease.

A summary of the data demonstrating a protective role for artificially-induced expression of Hsp70 or Hsp27 is given in Tables 1 and 2.

Concluding Remarks

Heat shock proteins and in particular Hsp70 and Hsp27 have emerged as potent inhibitors of apoptosis. They exert their effects by inhibiting protein aggregate formation and proteotoxicity, they block mitochondrial damage and they prevent caspase activation. Many neurodegenerative diseases display protein aggregation, oxidative stress and mitochondrial dysfunction, which suggests that Hsps could be promising therapeutic molecules. It is clear from the studies described above, that under conditions which cause apoptosis during neurodegenerative disease, depending on the cell type, animal strain and means of inducing over-expression of Hsp70 or Hsp27, protection can be achieved. However, before over-expression of heat shock proteins can be accepted as a means of reducing cell death, further studies will be required in order to define the criteria required for successful treatment.

Acknowledgements

This authors' work was supported financially by the Millennium Research Fund (NUI, Galway), Science Foundation of Ireland, Higher Education Authority of Ireland and the European Commission.

References

1. Hengartner MO. The biochemistry of apoptosis. Nature 2000; 407:770-776.
2. Samali A, Cai J, Zhivotovsky B et al. Presence of a preapoptotic complex of pro-caspase-3, Hsp60 and Hsp10 in the mitochondrial fraction of jurkat cells. EMBO J 1999; 18:2040-2048.
3. Fuentes-Prior P, Salvesen GS. The protein structures that shape caspase activity, specificity, activation and inhibition. Biochem J 2004; 384:201-232.
4. Philchenkov A. Caspases: Potential targets for regulating cell death. J Cell Mol Med 2004; 8:432-444.
5. Riedl SJ, Shi Y. Molecular mechanisms of caspase regulation during apoptosis. Nat Rev Mol Cell Biol 2004; 5:897-907.
6. Fischer U, Janicke RU, Schulze-Osthoff K. Many cuts to ruin: A comprehensive update of caspase substrates. Cell Death Differ 2003; 10:76-100.
7. Samali A, Gorman AM, Cotter TG. Apoptosis - The story so far. Experientia 1996; 52:933-941.
8. Thorburn A. Death receptor-induced cell killing. Cell Signal 2004; 16:139-144.
9. Wang S, El-Deiry WS. TRAIL and apoptosis induction by TNF-family death receptors. Oncogene 2003; 22:8628-8633.
10. Curtin JF, Cotter TG. Live and let die: Regulatory mechanisms in Fas-mediated apoptosis. Cell Signal 2003; 15:983-992.
11. Chang HY, Nishitoh H, Yang X et al. Activation of apoptosis signal-regulating kinase 1 (ASK1) by the adapter protein Daxx. Science 1998; 281:1860-1863.
12. Green DR, Reed JC. Mitochondria and apoptosis. Science 1998; 281:1309-1312.
13. Mishra NC, Kumar S. Apoptosis: A mitochondrial perspective on cell death. Indian J Exp Biol 2005; 43:25-34.
14. Polster BM, Fiskum G. Mitochondrial mechanisms of neural cell apoptosis. J Neurochem 2004; 90:1281-1289.
15. Zhou P, Chou J, Olea RS et al. Solution structure of Apaf-1 CARD and its interaction with caspase-9 CARD: A structural basis for specific adaptor/caspase interaction. Proc Natl Acad Sci USA 1999; 96:11265-11270.
16. Hill MM, Adrain C, Martin SJ. Portrait of a killer: The mitochondrial apoptosome emerges from the shadows. Mol Interv 2003; 3:19-26.
17. Nakagawa T, Zhu H, Morishima N et al. Caspase-12 mediates endoplasmic-reticulum-specific apoptosis and cytotoxicity by amyloid-beta. Nature 2000; 403:98-103.
18. Yoneda T, Imaizumi K, Oono K et al. Activation of caspase-12, an endoplastic reticulum (ER) resident caspase, through tumor necrosis factor receptor-associated factor 2-dependent mechanism in response to the ER stress. J Biol Chem 2001; 276:13935-13940.
19. Rao RV, Hermel E, Castro-Obregon S et al. Coupling endoplasmic reticulum stress to the cell death program. Mechanism of caspase activation. J Biol Chem 2001; 276:33869-33874.
20. Morishima N, Nakanishi K, Takenouchi H et al. An endoplasmic reticulum stress-specific caspase cascade in apoptosis. Cytochrome c-independent activation of caspase-9 by caspase-12. J Biol Chem 2002; 277:34287-34294.
21. Rutkowski DT, Kaufman RJ. A trip to the ER: Coping with stress. Trends Cell Biol 2004; 14:20-28.
22. Ma Y, Hendershot LM. The mammalian endoplasmic reticulum as a sensor for cellular stress. Cell Stress Chaperones 2002; 7:222-229.
23. Lee AH, Iwakoshi NN, Glimcher LH. XBP-1 regulates a subset of endoplasmic reticulum resident chaperone genes in the unfolded protein response. Mol Cell Biol 2003; 23:7448-7459.
24. Fischer H, Koenig U, Eckhart L et al. Human caspase 12 has acquired deleterious mutations. Biochem Biophys Res Commun 2002; 293:722-726.
25. Saleh M, Vaillancourt JP, Graham RK et al. Differential modulation of endotoxin responsiveness by human caspase-12 polymorphisms. Nature 2004; 429:75-79.
26. Hitomi J, Katayama T, Eguchi Y et al. Involvement of caspase-4 in endoplasmic reticulum stress-induced apoptosis and Abeta-induced cell death. J Cell Biol 2004; 165:347-356.
27. Liu X, Kim CN, Yang J et al. Induction of apoptotic program in cell-free extracts: Requirement for dATP and cytochrome c. Cell 1996; 86:147-157.
28. Lorenzo HK, Susin SA, Penninger J et al. Apoptosis inducing factor (AIF): A phylogenetically old, caspase-independent effector of cell death. Cell Death Differ 1999; 6:516-524.

29. Verhagen AM, Ekert PG, Pakusch M et al. Identification of DIABLO, a mammalian protein that promotes apoptosis by binding to and antagonizing IAP proteins. Cell 2000; 102:43-53.
30. Schafer P, Scholz SR, Gimadutdinow O et al. Structural and functional characterization of mitochondrial EndoG, a sugar nonspecific nuclease which plays an important role during apoptosis. J Mol Biol 2004; 338:217-228.
31. Melino G, Bernassola F, Ranalli M et al. p73 Induces apoptosis via PUMA transactivation and Bax mitochondrial translocation. J Biol Chem 2004; 279:8076-8083.
32. Xanthoudakis S, Roy S, Rasper D et al. Hsp60 accelerates the maturation of pro-caspase-3 by upstream activator proteases during apoptosis. EMBO J 1999; 18:2049-2056.
33. Li H, Zhu H, Xu CJ et al. Cleavage of BID by caspase 8 mediates the mitochondrial damage in the Fas pathway of apoptosis. Cell 1998; 94:491-501.
34. Luo X, Budihardjo I, Zou H et al. Bid, a Bcl2 interacting protein, mediates cytochrome c release from mitochondria in response to activation of cell surface death receptors. Cell 1998; 94:481-490.
35. Scaffidi C, Fulda S, Srinivasan A et al. Two CD95 (APO-1/Fas) signaling pathways. EMBO J 1998; 17:1675-1687.
36. Robertson JD, Gogvadze V, Zhivotovsky B et al. Distinct pathways for stimulation of cytochrome c release by etoposide. J Biol Chem 2000; 275:32438-32443.
37. Rich T, Allen RL, Wyllie AH. Defying death after DNA damage. Nature 2000; 407:777-783.
38. Willis S, Day CL, Hinds MG et al. The Bcl-2-regulated apoptotic pathway. J Cell Sci 2003; 116:4053-4056.
39. Saleh A, Srinivasula SM, Balkir L et al. Negative regulation of the Apaf-1 apoptosome by Hsp70. Nat Cell Biol 2000; 2:476-483.
40. Beere HM, Wolf BB, Cain K et al. Heat-shock protein 70 inhibits apoptosis by preventing recruitment of procaspase-9 to the Apaf-1 apoptosome. Nat Cell Biol 2000; 2:469-475.
41. Steel R, Doherty JP, Buzzard K et al. Hsp72 inhibits apoptosis upstream of the mitochondria and not through interactions with Apaf-1. J Biol Chem 2004; 279:51490-51499.
42. Creagh EM, Carmody RJ, Cotter TG. Heat shock protein 70 inhibits caspase-dependent and -independent apoptosis in Jurkat T cells. Exp Cell Res 2000; 257:58-66.
43. Ravagnan L, Gurbuxani S, Susin SA et al. Heat-shock protein 70 antagonizes apoptosis-inducing factor. Nat Cell Biol 2001; 3:839-843.
44. Clemens MJ. Translational control in virus-infected cells: Models for cellular stress responses. Semin Cell Dev Biol 2005; 16:13-20.
45. Gabai VL, Yaglom JA, Volloch V et al. Hsp72-mediated suppression of c-Jun N-terminal kinase is implicated in development of tolerance to caspase-independent cell death. Mol Cell Biol 2000; 20:6826-6836.
46. Mosser DD, Caron AW, Bourget L et al. Role of the human heat shock protein hsp70 in protection against stress-induced apoptosis. Mol Cell Biol 1997; 17:5317-5327.
47. Park J, Liu AY. JNK phosphorylates the HSF1 transcriptional activation domain: Role of JNK in the regulation of the heat shock response. J Cell Biochem 2001; 82:326-338.
48. Beere HM. "The stress of dying": The role of heat shock proteins in the regulation of apoptosis. J Cell Sci 2004; 117:2641-2651.
49. Jaattela M, Wissing D, Kokholm K et al. Hsp70 exerts its anti-apoptotic function downstream of caspase-3-like proteases. EMBO J 1998; 17:6124-6134.
50. Komarova EY, Afanasyeva EA, Bulatova MM et al. Downstream caspases are novel targets for the antiapoptotic activity of the molecular chaperone hsp70. Cell Stress Chaperones 2004; 9:265-275.
51. Park HS, Cho SG, Kim CK et al. Heat shock protein hsp72 is a negative regulator of apoptosis signal-regulating kinase 1. Mol Cell Biol 2002; 22:7721-7730.
52. Michels AA, Kanon B, Konings AW et al. Hsp70 and Hsp40 chaperone activities in the cytoplasm and the nucleus of mammalian cells. J Biol Chem 1997; 272:33283-33289.
53. Lu Z, Cyr DM. Protein folding activity of Hsp70 is modified differentially by the hsp40 cochaperones Sis1 and Ydj1. J Biol Chem 1998; 273:27824-27830.
54. Liu QL, Kishi H, Ohtsuka K et al. Heat shock protein 70 binds caspase-activated DNase and enhances its activity in TCR-stimulated T cells. Blood 2003; 102:1788-1796.
55. Gotoh T, Terada K, Oyadomari S et al. hsp70-DnaJ chaperone pair prevents nitric oxide- and CHOP-induced apoptosis by inhibiting translocation of Bax to mitochondria. Cell Death Differ 2004; 11:390-402.
56. Bruey JM, Ducasse C, Bonniaud P et al. Hsp27 negatively regulates cell death by interacting with cytochrome c. Nat Cell Biol 2000; 2:645-652.
57. Concannon CG, Orrenius S, Samali A. Hsp27 inhibits cytochrome c-mediated caspase activation by sequestering both pro-caspase-3 and cytochrome c. Gene Expr 2001; 9:195-201.

58. Garrido C, Bruey JM, Fromentin A et al. HSP27 inhibits cytochrome c-dependent activation of procaspase-9. FASEB J 1999; 13:2061-2070.
59. Paul C, Manero F, Gonin S et al. Hsp27 as a negative regulator of cytochrome C release. Mol Cell Biol 2002; 22:816-834.
60. Pandey P, Farber R, Nakazawa A et al. Hsp27 functions as a negative regulator of cytochrome c-dependent activation of procaspase-3. Oncogene 2000; 19:1975-1981.
61. Chauhan D, Li G, Hideshima T et al. Hsp27 inhibits release of mitochondrial protein Smac in multiple myeloma cells and confers dexamethasone resistance. Blood 2003; 102:3379-3386.
62. Charette SJ, Landry J. The interaction of HSP27 with Daxx identifies a potential regulatory role of HSP27 in Fas-induced apoptosis. Ann NY Acad Sci 2000; 926:126-131.
63. Thompson JE, Thompson CB. Putting the rap on Akt. J Clin Oncol 2004; 22:4217-4226.
64. Mearow KM, Dodge ME, Rahimtula M et al. Stress-mediated signaling in PC12 cells - The role of the small heat shock protein, Hsp27, and Akt in protecting cells from heat stress and nerve growth factor withdrawal. J Neurochem 2002; 83:452-462.
65. Rane MJ, Pan Y, Singh S et al. Heat shock protein 27 controls apoptosis by regulating Akt activation. J Biol Chem 2003; 278:27828-27835.
66. Parcellier A, Schmitt E, Gurbuxani S et al. HSP27 is a ubiquitin-binding protein involved in I-kappaBalpha proteasomal degradation. Mol Cell Biol 2003; 23:5790-5802.
67. Mehlen P, Schulze-Osthoff K, Arrigo AP. Small stress proteins as novel regulators of apoptosis. Heat shock protein 27 blocks Fas/APO-1- and staurosporine-induced cell death. J Biol Chem 1996; 271:16510-16514.
68. Mehlen P, Coronas V, Ljubic-Thibal V et al. Small stress protein Hsp27 accumulation during dopamine-mediated differentiation of rat olfactory neurons counteracts apoptosis. Cell Death Differ 1999; 6:227-233.
69. Arrigo AP, Virot S, Chaufour S et al. Hsp27 consolidates intracellular redox homeostasis by upholding glutathione in its reduced form and by decreasing iron intracellular levels. Antioxid Redox Signal 2005; 7:414-422.
70. Wyttenbach A, Sauvageot O, Carmichael J et al. Heat shock protein 27 prevents cellular polyglutamine toxicity and suppresses the increase of reactive oxygen species caused by huntingtin. Hum Mol Genet 2002; 11:1137-1151.
71. Preville X, Salvemini F, Giraud S et al. Mammalian small stress proteins protect against oxidative stress through their ability to increase glucose-6-phosphate dehydrogenase activity and by maintaining optimal cellular detoxifying machinery. Exp Cell Res 1999; 247:61-78.
72. Lavoie JN, Lambert H, Hickey E et al. Modulation of cellular thermoresistance and actin filament stability accompanies phosphorylation-induced changes in the oligomeric structure of heat shock protein 27. Mol Cell Biol 1995; 15:505-516.
73. Geum D, Son GH, Kim K. Phosphorylation-dependent cellular localization and thermoprotective role of heat shock protein 25 in hippocampal progenitor cells. J Biol Chem 2002; 277:19913-19921.
74. Van Why SK, Mann AS, Ardito T et al. Hsp27 associates with actin and limits injury in energy depleted renal epithelia. J Am Soc Nephrol 2003; 14:98-106.
75. Guay J, Lambert H, Gingras-Breton G et al. Regulation of actin filament dynamics by p38 map kinase-mediated phosphorylation of heat shock protein 27. J Cell Sci 1997; 110(Pt 3):357-368.
76. Huot J, Houle F, Rousseau S et al. SAPK2/p38-dependent F-actin reorganization regulates early membrane blebbing during stress-induced apoptosis. J Cell Biol 1998; 143:1361-1373.
77. Shimura H, Miura-Shimura Y, Kosik KS. Binding of tau to heat shock protein 27 leads to decreased concentration of hyperphosphorylated tau and enhanced cell survival. J Biol Chem 2004; 279:17957-17962.
78. Kosik KS, Shimura H. Phosphorylated tau and the neurodegenerative foldopathies. Biochim Biophys Acta 2005; 1739:298-310.
79. Lassmann H, Bancher C, Breitschopf H et al. Cell death in Alzheimer's disease evaluated by DNA fragmentation in situ. Acta Neuropathol (Berl) 1995; 89:35-41.
80. Smale G, Nichols NR, Brady DR et al. Evidence for apoptotic cell death in Alzheimer's disease. Exp Neurol 1995; 133:225-230.
81. Dragunow M, Preston K, Dodd J et al. Clusterin accumulates in dying neurons following status epilepticus. Brain Res Mol Brain Res 1995; 32:279-290.
82. Sugaya K, Reeves M, McKinney M. Topographic associations between DNA fragmentation and Alzheimer's disease neuropathology in the hippocampus. Neurochem Int 1997; 31:275-281.
83. Thomas LB, Gates DJ, Richfield EK et al. DNA end labeling (TUNEL) in Huntington's disease and other neuropathological conditions. Exp Neurol 1995; 133:265-272.
84. Vis JC, Schipper E, de Boer-van Huizen RT et al. Expression pattern of apoptosis-related markers in Huntington's disease. Acta Neuropathol (Berl) 2005.

85. Tatton WG, Chalmers-Redman R, Brown D et al. Apoptosis in Parkinson's disease: Signals for neuronal degradation. Ann Neurol 2003; 53(Suppl 3):S61-70, (discussion S70-62).
86. Tatton NA. Increased caspase 3 and Bax immunoreactivity accompany nuclear GAPDH translocation and neuronal apoptosis in Parkinson's disease. Exp Neurol 2000; 166:29-43.
87. de la Monte SM, Sohn YK, Ganju N et al. P53- and CD95-associated apoptosis in neurodegenerative diseases. Lab Invest 1998; 78:401-411.
88. Hartmann A, Hunot S, Michel PP et al. Caspase-3: A vulnerability factor and final effector in apoptotic death of dopaminergic neurons in Parkinson's disease. Proc Natl Acad Sci USA 2000; 97:2875-2880.
89. Hartmann A, Troadec JD, Hunot S et al. Caspase-8 is an effector in apoptotic death of dopaminergic neurons in Parkinson's disease, but pathway inhibition results in neuronal necrosis. J Neurosci 2001; 21:2247-2255.
90. Giffard RG, Yenari MA. Many mechanisms for hsp70 protection from cerebral ischemia. J Neurosurg Anesthesiol 2004; 16:53-61.
91. Matsumori Y, Hong SM, Aoyama K et al. Hsp70 overexpression sequesters AIF and reduces neonatal hypoxic/ischemic brain injury. J Cereb Blood Flow Metab 2005.
92. Kelly S, Yenari MA. Neuroprotection: Heat shock proteins. Curr Med Res Opin 2002; 18(Suppl 2):s55-60.
93. Amin V, Cumming DV, Latchman DS. Over-expression of heat shock protein 70 protects neuronal cells against both thermal and ischaemic stress but with different efficiencies. Neurosci Lett 1996; 206:45-48.
94. Kwong JM, Lam TT, Caprioli J. Hyperthermic preconditioning protects retinal neurons from N-methyl-D-aspartate (NMDA)-induced apoptosis in rat. Brain Res 2003; 970:119-130.
95. Xu L, Lee JE, Giffard RG. Overexpression of bcl-2, bcl-XL or hsp70 in murine cortical astrocytes reduces injury of cocultured neurons. Neurosci Lett 1999; 277:193-197.
96. Whitlock NA, Lindsey K, Agarwal N et al. Heat shock protein 27 delays Ca2+-induced cell death in a caspase-dependent and -independent manner in rat retinal ganglion cells. Invest Ophthalmol Vis Sci 2005; 46:1085-1091.
97. Dong Z, Wolfer DP, Lipp HP et al. Hsp70 gene transfer by adeno-associated virus inhibits MPTP-induced nigrostriatal degeneration in the mouse model of Parkinson disease. Mol Ther 2005; 11:80-88.
98. Zhou Y, Gu G, Goodlett DR et al. Analysis of alpha-synuclein-associated proteins by quantitative proteomics. J Biol Chem 2004; 279:39155-39164.
99. Quigney DJ, Gorman AM, Samali A. Heat shock protects PC12 cells against MPP^+ toxicity. Brain Res 2003; 993:133-139.
100. Gorman AM, Szegezdi E, Quigney DJ et al. Hsp27 inhibits 6-hydroxydopamine-induced cytochrome c release and apoptosis in PC12 cells. Biochem Biophys Res Commun 2005; 327:801-810.
101. Dedmon MM, Christodoulou J, Wilson MR et al. Heat shock protein 70 inhibits {alpha}-synuclein fibril formation via preferential binding to prefibrillar species. J Biol Chem 2005; 280:14733-14740.
102. Klucken J, Shin Y, Masliah E et al. Hsp70 Reduces alpha-synuclein aggregation and toxicity. J Biol Chem 2004; 279:25497-25502.
103. Zourlidou A, Payne Smith MD, Latchman DS. HSP27 but not HSP70 has a potent protective effect against alpha-synuclein-induced cell death in mammalian neuronal cells. J Neurochem 2004; 88:1439-1448.
104. Magrane J, Smith RC, Walsh K et al. Heat shock protein 70 participates in the neuroprotective response to intracellularly expressed beta-amyloid in neurons. J Neurosci 2004; 24:1700-1706.
105. Zhang Y, Champagne N, Beitel LK et al. Estrogen and androgen protection of human neurons against intracellular amyloid beta1-42 toxicity through heat shock protein 70. J Neurosci 2004; 24:5315-5321.
106. Rossler OG, Bauer I, Chung HY et al. Glutamate-induced cell death of immortalized murine hippocampal neurons: Neuroprotective activity of heme oxygenase-1, heat shock protein 70, and sodium selenite. Neurosci Lett 2004; 362:253-257.
107. Lee JE, Yenari MA, Sun GH et al. Differential neuroprotection from human heat shock protein 70 overexpression in vitro and in vivo models of ischemia and ischemia-like conditions. Exp Neurol 2001; 170:129-139.
108. Rajdev S, Hara K, Kokubo Y et al. Mice overexpressing rat heat shock protein 70 are protected against cerebral infarction. Ann Neurol 2000; 47:782-791.
109. Plumier JC, Krueger AM, Currie RW et al. Transgenic mice expressing the human inducible Hsp70 have hippocampal neurons resistant to ischemic injury. Cell Stress Chaperones 1997; 2:162-167.

CHAPTER 6

Assembly of Protein Aggregates in Neurodegeneration:

Mechanisms Linking the Ubiquitin/Proteasome Pathway and Chaperones

Sha-Ron Pierre, Vita Vernace, Zhiyou Wang and Maria E. Figueiredo-Pereira*

Abstract

In recent years, it has become increasingly evident that the majority of neurodegenerative disorders is associated with the aggregation and deposition of proteins in inclusion bodies. To avoid this abnormal deposition of proteins the cells recruit molecular chaperones to suppress aggregation and the ubiquitin/proteasome pathway (UPP) to remove the aggregate-prone proteins. The UPP is the major nonlysosomal degradation pathway for intracellular proteins. UPP impairment and/or its overload are likely to be major contributors to the aggregation of ubiquitinated proteins detected in most neuronal inclusion bodies. The mechanisms leading to the formation of inclusion bodies are not well defined. In this chapter, we discuss cellular strategies to deliver substrates to the UPP and their potential contribution to the development of intracellular proteins aggregates. In the future, a better understanding of the steps leading to protein aggregation and their deposition in inclusion bodies is likely to provide opportunities for effective therapeutic interventions.

Introduction

Proteolysis is an important cellular event involving tightly regulated removal of unwanted proteins and retention of those that are essential. In addition to its function in normal protein degradation, the ubiquitin/proteasome pathway (UPP) plays a critical role in the quality control process. The UPP eliminates mutated or abnormally modified proteins by degradation to prevent their accumulation as aggregates that often form intracellular inclusion bodies.

Intracellular inclusion bodies containing ubiquitinated proteins are detected in a variety of degenerative diseases. These diseases range from neurological disorders, such as Alzheimer's disease, Parkinson's disease, amyotrophic lateral sclerosis, Huntington's disease and spinocerebellar ataxias, to liver diseases, such as Wilson's disease and alcoholic hepatitis, to name a few.[1] Whether these inclusion bodies are pathogenic or represent a coping mechanism to prolong survival of the affected cells, such as neurons and hepatocytes, is a hotly debated issue. The abnormal protein aggregates, however, are indicative of a malfunction of the process of protein turnover since they are not prevalent in healthy individuals.

*Corresponding Author: Maria E. Figueiredo-Pereira—Department of Biological Sciences, Hunter College of CUNY, New York, New York, U.S.A. Email: pereira@genectr.hunter.cuny.edu

Heat Shock Proteins in Neural Cells, edited by Christiane Richter-Landsberg.

The major structural components of inclusion bodies vary from cell type to cell type. For example, microtubule associated *tau* proteins are found in cortical neurofibrillary tangles, α-synuclein in dopaminergic Lewy bodies and huntingtin in nuclear inclusions in the striatum (reviewed in ref. 2). However, most of these protein deposits contain ubiquitinated proteins. The massive neuronal deposition of ubiquitinated proteins in inclusion bodies in neurodegenerative disorders implicates the UPP in their formation. It is likely that these protein aggregates develop when the capacity of the proteasome as well as that of molecular chaperones is exceeded by the production of aggregation-prone misfolded proteins. The mechanisms leading to the formation of these protein aggregates remain to be fully characterized. In an effort to provide a general overview of the current information available on this topic, we will address the following issues: (1) Protein degradation by the UPP; (2) Potential mechanisms for the formation of ubiquitin-protein aggregates; (3) Subcellular distribution of ubiquitin-protein aggregates. This review will end with (4) Conclusions. Please note that this review is not intended to be comprehensive and we apologize to the authors whose work is not mentioned here.

Protein Degradation by the UPP

The proteolytic mechanism of the UPP has a broad specificity, cleaving peptide bonds after basic, acidic and hydrophobic amino acids. This pathway requires most proteins to be tagged by ubiquitin for proteasomal targeting. However, there is a small set of proteins, such as ornithine decarboxylase, that is targeted for proteasomal degradation by ubiquitin-independent means (reviewed in ref. 3). Besides ubiquitination, recent studies established that folded proteins require an additional targeting signal for efficient proteasomal degradation.[4] Indeed, an unstructured initiation site on the protein substrate significantly accelerates its degradation by the proteasome.

In general, proteolysis by the UPP involves two major steps: ubiquitination followed by degradation. A de-ubiquitination step also plays important roles in this pathway as it edits the ubiquitination state of the different substrates and removes the ubiquitin tag for recycling.

Ubiquitination/De-Ubiquitination

Ubiquitin (Ub) is a small protein of 76 amino acids and it can form polyubiquitin chains. To target proteins for proteasomal degradation, polyubiquitin chains are formed by the successive attachment of monomers by an isopeptide bond formed between the side chain of Lys48 in one ubiquitin and the carboxyl group of the C-terminal Gly76 of a neighboring Ub. Attachment of polyubiquitin chains to lysine residues on a protein results in at least a 10-fold increase in its degradation rate.[5] Polyubiquitin chains with linkages involving lysine residues other than Lys48 on ubiquitin, were found to play distinct roles. These alternate polyubiquitin chains play a role in DNA repair, activation of NF-κB, polysome stability and endocytosis, to name a few.[6]

Polyubiquitination of proteins is a complex process involving four steps (Fig. 1). The first step involves the formation of a high-energy thioester bond between Ub and a ubiquitin-activating enzyme (E1) in a reaction that requires ATP hydrolysis. In the second step a thioester bond is formed between the activated ubiquitin and ubiquitin-conjugating enzymes (E2). In the third step the carboxyl terminal of Ub is covalently attached to the protein substrate. This reaction is mediated by ubiquitin ligases (E3), which confer substrate specificity to the UPP. In general, the first ubiquitin conjugated to a protein is attached to the ε-amino group of substrate lysines via an isopeptide bond. However, recent findings demonstrated that ubiquitin can also be conjugated to the α-amino group of the N-terminal residue (reviewed in ref. 7) or to intramolecular cysteines of the substrate.[8] In some cases, ubiquitin can be transferred directly to the protein substrate by ubiquitin-conjugating enzymes (E2). In the fourth step, multiubiquitin chains are assembled by a family of ubiquitination factors (E4) that produces longer Ub-chains.[9]

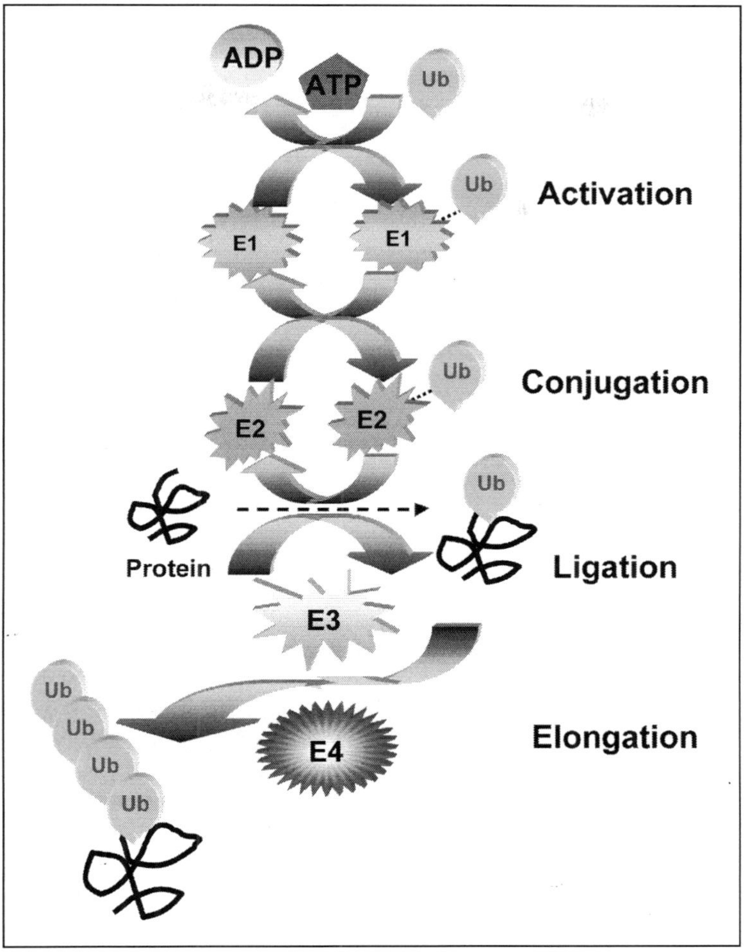

Figure 1. Protein ubiquitination. First, a high-energy thioester bond is formed between ubiquitin (Ub) and a ubiquitin-activating enzyme (E1). This reaction requires ATP hydrolysis. Secondly, the activated ubiquitin is transferred to a ubiquitin conjugating enzyme (E2). Thirdly, the activated ubiquitin is ligated to the protein substrate by a ubiquitin ligase (E3). Lastly, the ubiquitin chain is elongated by an ubiquitin-chain elongating factor (E4), which drives the assembly of the polyubiquitin chain.

While the number of distinct E1 enzymes is small, there are many E2 and E3 enzymes, indicating that this pathway operates through selective proteolysis (reviewed in ref. 10). At least 50 E2s were identified in humans.[11] They share a common 150-amino acid catalytic core, whereas each subfamily possesses affinity for a different class of E3 enzymes. E3s recognize specific protein substrates for ubiquitination and coordinate the ligation of ubiquitin to the substrate. The human genome contains at least 1,000 different E3s, providing high specificity to the UPP.[7] Mechanistically, two classes of E3s have been described (reviewed in ref. 11). One class, the HECT-domain E3s,[12] have a conserved Cys residue that participates in the transfer of activated ubiquitin to a target protein. The HECT-E3s form ubiquitin-thioester intermediates and ubiquitinate substrates directly. The second class of E3s is not catalytic. Instead, these E3s stabilize a molecular scaffold for E2 interaction with the substrate. The E2 will then ligate ubiquitin to the substrate. Some of these E3s consist of just one subunit (RING finger domain E3s) and others are multi-subunit complexes (SCF-type E3, VBC-Cul2 E3, anaphase promoting complex, to name a few).

The assembly of polyubiquitin chains can be a processive reaction requiring E1, E2 and E3 enzymes. In many instances, another enzyme is recruited that provides for a more efficient polyubiquitin chain assembly. These polyubiquitin elongation factors are collectively known as E4s (reviewed in ref. 13). So far, three types of E4s have been identified. The first type includes the U-box-containing E4s. The U-box is a 70-residue domain structurally related to the E3 RING-finger domain. One example of this E4 type is CHIP (C-terminus of Hsc70-interacting protein) a protein that is particularly relevant to neurodegeneration. Upon CHIP interaction with HSP90 and HSP70 it acts as an E3 ligase leading to the ubiquitination and proteasomal degradation of misfolded proteins that are chaperone-bound (reviewed in refs. 13,14). CHIP in collaboration with the E3 ligase parkin, was also shown to act as an E4 elongation factor for the unfolded Pael receptor.[15] The second type of E4s includes nonU-box E4s for which p300 is an example. This transcriptional cofactor was shown to attach polyubiquitin chains to p53 previously monoubiquitinated by the E3 ligase MDM2 (reviewed in ref. 13). Finally, the third type of E4s, so far only identified in *C. elegans*, consists of an E3-E4 polyubiquitinating complex (reviewed in ref. 13). In this case, two E3s, CHN-1 and UFD-2 act in concert as an E4, thus polyubiquitinating the myosin chaperone UNC-45.

Ubiquitin is removed from ubiquitinated proteins by de-ubiquitinating enzymes, which also disassemble polyubiquitin chains. More than 70 genes encoding for de-ubiquitinating enzymes have been identified in humans.[16] There are two major classes of de-ubiquitinating enzymes. The first class, the ubiquitin carboxyl-terminal hydrolases (UCHs), removes small amides, esters, peptides and small proteins at the carboxyl terminus of ubiquitin. The second class, the ubiquitin-specific processing proteases (UBPs), disassembles the polyubiquitin chains and edits the ubiquitination state of proteins.[17]

Proteasome Degradation

One of the functions of covalently binding Ub to proteins is to mark them for degradation by the 26S proteasome (Fig. 2), a multicomponent enzymatic complex with a molecular mass of approximately 2,000 kDa.[18] The 26S proteasome includes two major particles: a 20S particle, known as the 20S proteasome, which is the catalytic core, and a 19S particle, known as PA700, which is the regulatory component.

The 20S particle is composed of 28 subunits arranged in four heptameric-stacked rings forming a cylindrical structure with a hollow center in which proteolysis takes place.[19] The 20S proteasome hydrolyzes most peptide bonds present in a protein,[20] and its rate of hydrolysis is influenced by its subunit composition.[21] Assembly of this particle from precursor subunits is a complex process and requires the assistance of a short-lived chaperone.[22]

The 19S particle contains at least 15 subunits, including ATPases, a de-ubiquitinating enzyme and polyubiquitin-binding subunits. It confers ubiquitin/ATP-dependency to proteolysis by the 26S proteasome.[18] The subunits in the 19S particle are distributed into a lid and base arrangement, with the lid required for ubiquitin/ATP-dependent peptide bond hydrolysis.[23] The base containing the ATPases exhibits chaperone-like activity.[24]

Association between the two particles in the cell is a dynamic process and requires ATP-hydrolysis. The 20S proteasome can associate with other regulatory members, such as PA28, but this combination is not known to regulate the degradation of polyubiquitinated proteins.[18]

The 26S proteasome is found in the cytoplasm next to intermediate filaments of the cytoskeleton.[25] It also resides in the nucleus and in association with the cytoplasmic side of the ER membrane.[26,27] Localization studies with fluorescently labeled subunits of the 20S and 19S particles demonstrated that proteasomal proteolysis occurs mainly at the nuclear

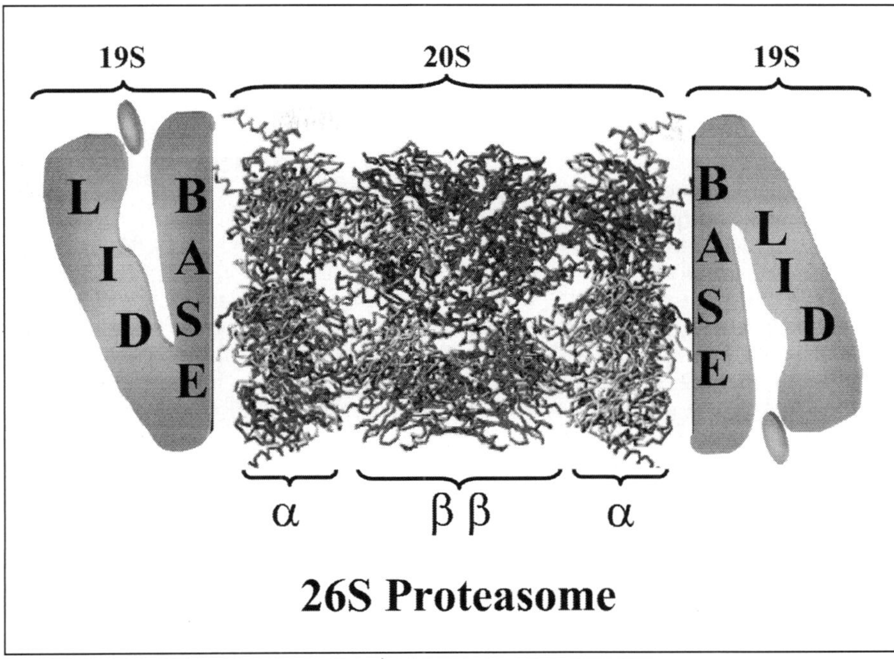

Figure 2. The 26S proteasome. Its two major particles, the 20S particle (20S proteasome) which is the catalytic core, and the 19S particle (PA700) which is the regulatory component, require ATP hydrolysis to assemble into the 26S proteasome. The crystal structure of the 20S proteasome was obtained from 1ryp.pdb.[19]

envelope/rough ER site.[28] An important function of such proteolysis is to eliminate abnormal secretory proteins residing in an ER/preGolgi compartment.[29] Functionally inefficient, misfolded or unassembled ER proteins leave this intracellular compartment by retrograde transport through the Sec61 translocation channel. They are ubiquitinated by ubiquitin-conjugating enzymes associated with the cytosolic side of the ER membrane and then degraded by the cytosolic 26S proteasome.[29] Although this ER degradation pathway appears to be nonessential for viability, its importance is underscored by its evolutionary preservation "despite strong negative selection" since disruption of this mechanism seems to be associated with many disease states.[29]

Potential Mechanisms for the Formation of Ubiquitin-Protein Aggregates

The role of inclusion bodies in the progression of neurodegeneration is unknown.[30] On the one hand inclusions may be beneficial and result from an attempt of the cell to isolate a subclass of ubiquitinated proteins that are not effectively degraded. On the other hand, the inclusions may impede normal cell function contributing to cell death. The size and abundance of the inclusions maybe critical determinants of their toxicity.[31] Accordingly, small inclusions may be protective while expanded inclusions may confer fatal effects that can contribute to neuronal cell damage.

Several mechanisms leading to the formation of intracellular aggregates of ubiquitinated proteins have been proposed. We will discuss four of them that may be relevant to the development of inclusion bodies in neurodegenerative disorders.

UBA/UBL "Noncanonical" Chaperones

The first mechanism involves "noncanonical" chaperones that deliver ubiquitinated proteins to the 26S proteasome. These potential shuttles for polyubiquitinated proteins (Fig. 3) contain an ubiquitin-like (UBL) domain at the N-terminus and at least one ubiquitin-associated (UBA) domain at the C-terminus (reviewed in refs. 32,33). The UBL domain is known to interact with the 19S particle of the proteasome, in particular with the subunit S5a/Rpn10.[34] The UBA domain noncovalently binds polyubiquitin chains up to 300-times more tightly than mono-ubiquitin (reviewed in refs. 35,36).

In the context of neurodegeneration one of the UBL/UBA proteins that is best characterized is the sequestosome1, also known as p62, first identified in human tissues by Shin and colleagues.[37] The sequestosome1/p62 was detected in ubiquitin-containing intraneuronal and intraglial inclusions in a variety of neurodegenerative disorders[38-41] as well as in hepatocyte Mallory bodies associated with alcoholism.[42] These findings suggest that the sequestosome1/p62 is relevant to the biogenesis of inclusions containing ubiquitinated proteins. Furthermore, sequestosome1/p62 expression in neuronal cells is induced by serum withdrawal conditions that trigger apoptosis, by expression of expanded pathologic polyglutamine repeats[43] and by proteasome inhibitors[44,45] as well as prostaglandins of the J2 series.[46] Due to its high affinity for polyubiquitin chains, sequestosome1/p62 was suggested to serve as a receptor for binding and storing ubiquitinated proteins.[47] Recent studies with isolated HEK cells transfected with full-length or truncated sequestosome forms, indicate that sequestosome1/p62 may act as a shuttle that delivers polyubiquitinated proteins to the proteasome.[34] Interestingly, these studies suggest that sequestosome1/p62 can bind polyubiquitin chains through the C-terminal UBA domain and the proteasome through its AID (acidic interaction domain), which is closer to the N-terminus. The AID is proposed to be structurally similar to ubiquitin-like (UBL) domains known to interact with the proteasome. Due to its binding versatility, the sequestosome1/p62 may play an important role as a scaffold and/or shuttle molecule storing polyubiquitinated proteins and delivering them to the proteasome in a regulated manner.

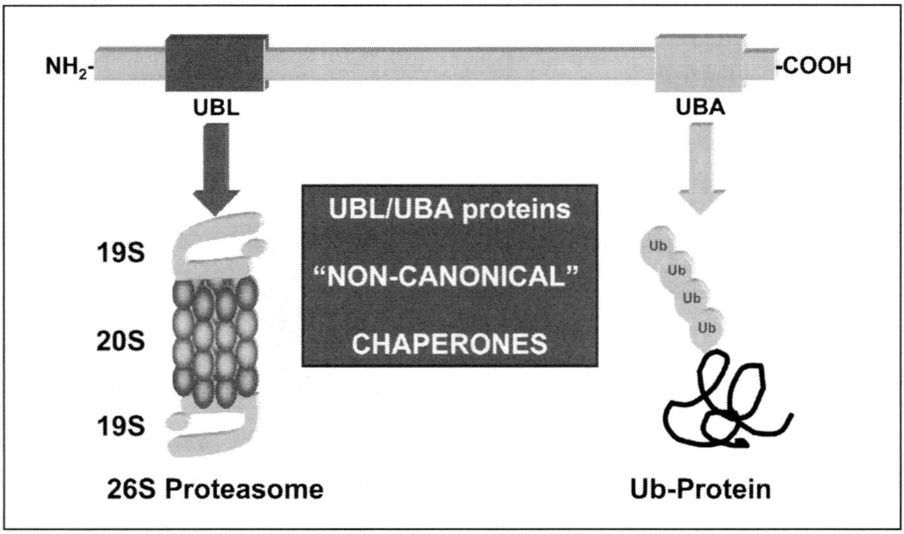

Figure 3. UBL/UBA proteins. These "noncanonical" chaperones exhibit a UBL (ubiquitin-like) domain at the N-terminus and at least one UBA (ubiquitin-associated) domain at the C-terminus. These proteins are thought to shuttle polyubiquitinated proteins to the proteasome. The UBL domain is a "receptor" for particular subunits of the 19S particle of the 26S proteasome and the UBA domain is a "receptor" for polyubiquitin chains.

The discovery of proteins with UBL/UBA domains supports the view that translocation of ubiquitinated proteins to the proteasome is a regulated, not a stochastic process. This "shuttling" mechanism is likely to include not only UBL/UBA proteins but also other factors. For example, one of the proteins (ataxin 3) that cause polyglutamine-neurodegenerative diseases, exhibits the ability to bind polyubiquitin chains and de-ubiquitinating activity as well.[48] Both ataxin 3 activities are required for the formation of protein aggregates.[49] Another polyglutamine-neurodegenerative protein, ataxin 1, disrupts the interaction of the UBA/UBL protein A1UP with the 26S proteasome.[50] To make matters more complex, some UBA/UBL proteins in yeast were shown to inhibit the degradation of model polyubiquitinated substrates.[51] Delivery of ubiquitinated proteins to the proteasome is thus still poorly understood and remains an exciting future challenge.

The sequestration of polyubiquitinated proteins by the sequestosome1/p62 or other UBL/UBA proteins could act as a cellular defense mechanism. Cellular stress conditions yielding large quantities of misfolded/polyubiquitinated proteins would benefit from their sequestration by these "noncanonical" chaperones. Up-regulation of these "noncanonical" chaperones triggered by increased levels of misfolded proteins would prevent them from blocking the proteasome. The latter would provide an efficient mechanism to regulate substrate access to the proteasome and prevent its shutdown by excessive amounts of substrates. These types of intracellular storage aggregates would thus prevent cell damage and promote cell survival. However, under extreme stress conditions the overabundance of misfolded proteins could jeopardize the delivery process and activate cellular death pathways.

Aggresomes

The second mechanism leading to the formation of aggregates of ubiquitinated proteins involves aggresomes, first described by Kopito's group.[52] Aggresomes are thought to be deposition sites for ubiquitinated proteins that escape degradation by the UPP (Fig. 4). They are cytoplasmic inclusions found in an indentation of the nuclear envelope that colocalizes with centrosome/MTOC (microtubule organizing center) markers (reviewed in ref. 53). Notably, previous studies demonstrated that the microtubule "minus end" motor activities directed by dynein are necessary for the formation of aggresomes,[54] which also contain dynein.[55] Dyneins direct intracellular cargos toward the cell center (and nucleus) where the microtubule "minus ends" are clustered at the MTOC.[56] While some studies suggest that the retrograde transport of ubiquitin protein aggregates to centrosomes is dependent on microtubule integrity,[52] others indicate that this process may not require intact microtubules.[57]

Aggresomes are composed of insoluble material and are associated with high levels of 26S proteasomes as well as with de-ubiquitination activity.[57] Some of the ubiquitinated proteins shown to be deposited in aggresomes resulted from either overexpression of mutant cystic fibrosis transmembrane conductor regulator or presenilin 1 and/or from impaired protein degradation induced by treating cells with proteasome inhibitors.[52,54,58]

Currently, it is not clear if the mechanisms shown to participate in the formation of aggresomes occur under homeostatic and/or nonhomeostatic conditions. Aggresome-formation was demonstrated to occur mainly as a result of overexpression of mutant proteins in cells often treated with proteasome inhibitors. Therefore, aggresome formation has not been demonstrated in vivo.

DALIS

A third mechanism leading to the formation of aggregates containing ubiquitin-conjugates was identified in maturing dendritic cells by Pierre's group.[59] These transient (reversible) aggregates of ubiquitinated proteins (Fig. 5) are distinct from aggresomes. They are not colocalized with MTOC markers, they don't exhibit vimentin cages and they do not inhibit the proteasome.[60] Upon stimulation with pro-inflammatory agents followed by treatment with protein damaging agents, the ensuing defective ribosomal products (DRiPs) are sorted into large cytosolic aggregates known as dendritic cell aggresome-like induced structures (DALIS). DALIS contain many components of the ubiquitination machinery, including E1, E2s and E3s. When DRiPs

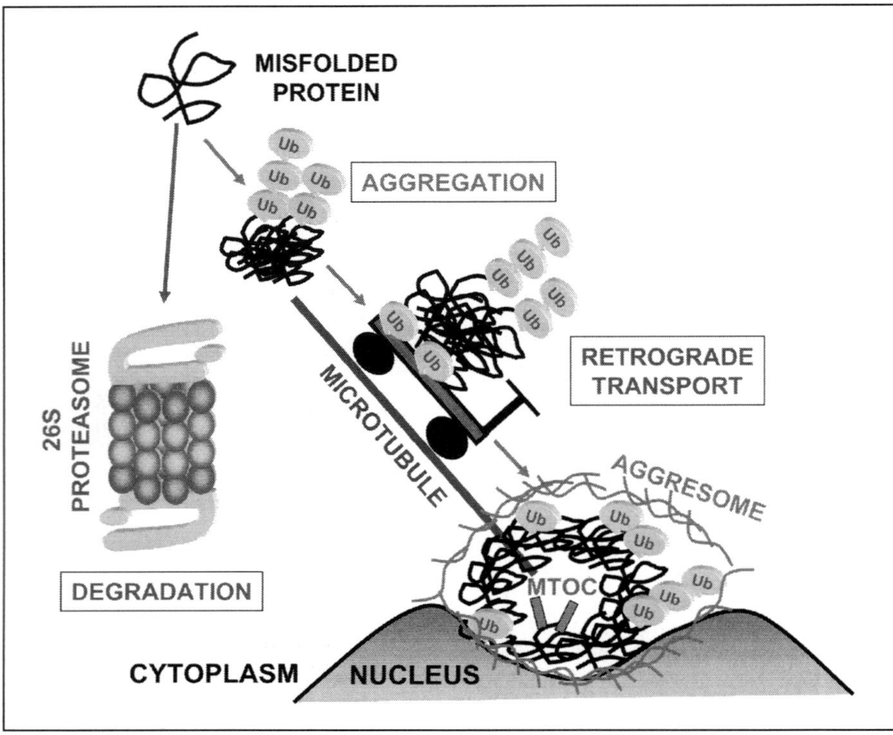

Figure 4. Aggresomes. Aggresomes result from the microtubule-dependent retrograde transport of small aggregates of ubiquitinated proteins to the area of the microtubule-organizing center (MTOC). The aggregates are shuttled by dynein complexes on microtubule tracks. Vimentin cages surround the aggresomes.

are formed they are rapidly sequestered into DALIS where they are eventually ubiquitinated. This mechanism allows dendritic cells to regulate the degradation rate of DRiPs, an ability that is pivotal to their immune functions.[60] DALIS, are thus thought to act as antigen storage structures allowing for the regulated degradation of proteins during infection. Recent studies demonstrated that DALIS-formation is also induced in dendritic cells by heat shock[61] and in macrophages by microbial products.[62]

DALIS-formation requires continuous protein synthesis.[60] It is thus tempting to speculate that up-regulation of UBL/UBA proteins may play a role in the sequestration of DRiPs into DALIS. However, the mechanisms involved in sorting DRiPs into DALIS and how DALIS are disassembled, remain to be characterized. Furthermore, if a DALIS-like sorting of defective and/or mutated proteins occurs in neuronal cells leading initially to transient aggregates, remains to be established. It is tempting to propose that overproduction of neuronal defective proteins and/or UPP impairment, disturb the DALIS-sorting mechanism resulting in permanent nonreversible aggregates such as the ones detected in neurodegeneration.

Russell Bodies

The fourth and final mechanism relevant to the formation of protein aggregates (in this case, of nonubiquitinated proteins) is related to the endoplasmic reticulum (ER) (reviewed in ref. 63). The ER aggregates known as Russell bodies were first discovered more than 100 years ago in cancer cells by Russell.[64] More recently, Russell bodies were described as being swollen ER cisternae that contain insoluble aggregates of mutant immunoglobulin.[65] They are

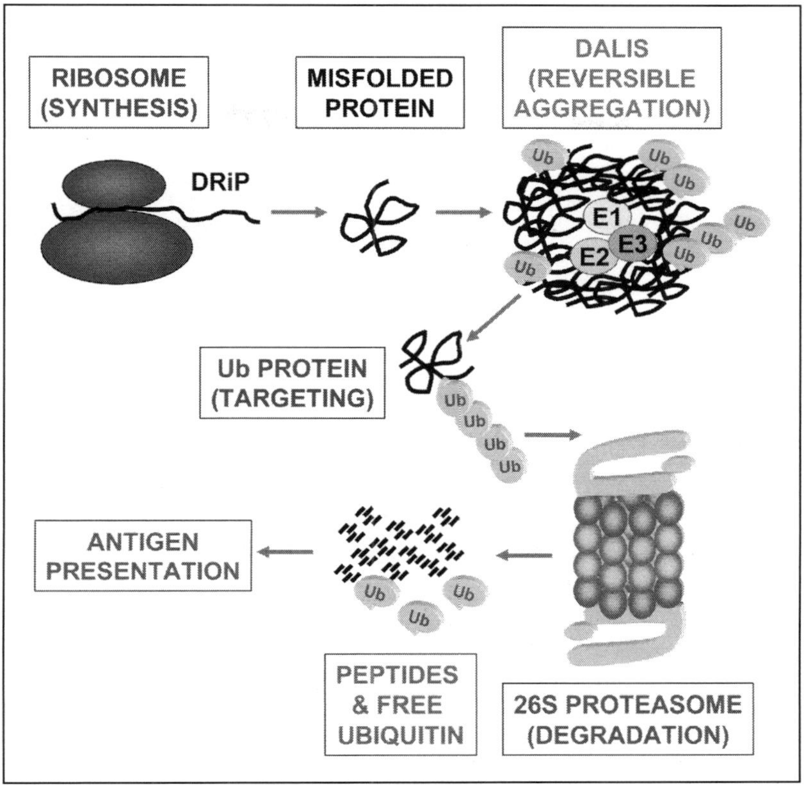

Figure 5. DALIS. Defective ribosomal products (DRiPs) are considered a main source for MHC class I-restricted antigenic peptides. DRiPs are incorporated into dendritic cell aggresome-like induced structures (DALIS) where they are ubiquitinated and transiently stored for processing by the UPP, for antigen presentation. Limiting DRiP degradation and storing them in DALIS allows dendritic cells to mature and control MHC class I loading and antigen presentation.

postulated to form as a cellular attempt to compartmentalize abnormal ER proteins that cannot escape into the cytoplasm to be degraded (Fig. 6). Instead, they are sequestered into ER subcompartments to prevent blockage of the normal secretory pathway.[65]

No Russell bodies were yet identified in neurons. In addition, Russell bodies do not contain ubiquitinated proteins as the ubiquitination machinery is absent from the ER lumen. Thus no correlation has been established between this ER sorting mechanism and the aggregation of ubiquitinated proteins detected in neurodegenerative diseases. However, this type of ER sequestration response may be relevant to the aggregation of proteins that, although nonubiquitinated, are residents of the secretory pathway and relevant to neurodegeneration.

Subcellular Distribution of Ubiquitin-Protein Aggregates

The intracellular aggregates containing ubiquitinated proteins detected in neurodegenerative disorders, vary in their subcellular distribution. Some are detected in the cytoplasm, such as in Parkinson's disease and others in the nucleus, such as in Huntington's disease. The cause of this differential subcellular aggregate distribution is not clear. An investigation of the nuclear diffusion limit in mammalian cells, including primary neuronal cells, established that large molecules (molecular masses above 70 kDa) cannot freely diffuse into nuclei of intact, healthy

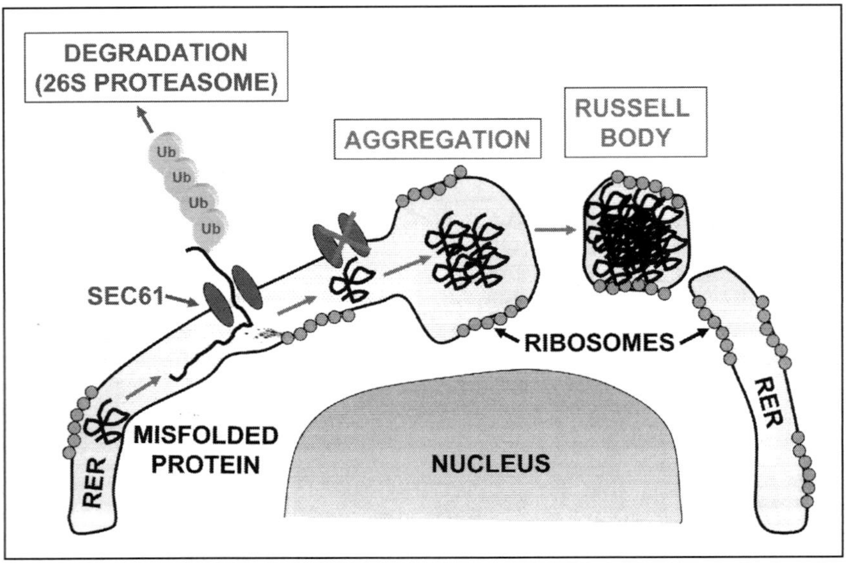

Figure 6. Russell bodies originate from RER (rough endoplasmic reticulum) subcompartments containing nonsecreted proteins that escape intracellular proteolysis. These nondegradable proteins fail to be translocated into the cytoplasm through the Sec61-based channel. Instead they accumulate and aggregate in RER subcompartments and form Russell bodies. These structures may represent a general cellular response to the aggregation of nondegradable ER proteins.

cells.[66] It is probable that many of the ubiquitin protein conjugates detected in inclusion bodies have molecular masses above 70kDa as estimated by western blot analysis. It is unlikely that these high molecular mass ubiquitin conjugates passively diffuse from the cytoplasm into the nucleus and vice-versa. The large aggregates could only passively enter the nucleus if the nuclear membrane was disrupted. Accordingly, nuclear migration of full-length mutant huntingtin can only occur upon deterioration of the nuclear membrane.[66] Wild type huntingtin is a cytoplasmic protein with a molecular mass of ~350kDa. Neither wild type nor mutant huntingtin have a nuclear targeting signal and thus cannot be actively transported across the nuclear membrane.[66] Interestingly, the subcellular distribution of transfected GFP (~30kDa) fused to full length or truncated forms of wild type or mutant huntingtin is not homogeneous. These fusion proteins can accumulate in the cytoplasm, nucleus or both, depending on the size of the fusion protein, on the nuclear diffusion limit of the specific transfected cells and on their nuclear membrane integrity.[66] We conclude that aggregate size, nuclear diffusion limit and nuclear membrane integrity are some of the factors that determine the subcellular distribution of protein aggregates.

Concluding Remarks

Evidence implicating molecular chaperones in neurodegenerative diseases is more compelling than that for the UPP, mainly because a direct link between UPP impairment and neurodegeneration is next to impossible to prove in postmortem human tissue. However, studies with human disease tissue indicate that disturbances of protein degradation by the UPP must have catastrophic consequences and play a critical role in neurodegeneration. It is clear that a full-cooperation between the proteolytic and chaperone systems is required to prevent the development of potentially toxic protein aggregates. The UPP is recruited for removal of proteins that are most likely modified, misfolded, ubiquitin-tagged and that escape appropriate refolding by molecular chaperones. However, the overwhelming production of misfolded proteins alone or in conjunction with UPP impairment is likely to lead to the formation of protein aggregates most of them containing ubiquitinated proteins. We are just beginning to decipher the intracellular mechanisms (some of them described above) that contribute to protein aggregation. A better understanding of this process will most certainly lead to new routes for therapy.

Acknowledgements

This work was supported by National Institutes of Health Grants NS34018 (M.E. Figueiredo-Pereira) as well as GM60654 (NIGMS) and RR03037 (NIGMS/RCMI core facility grant) both to Hunter College of CUNY.

References

1. Lowe J, Blanchard A, Morrell K et al. Ubiquitin is a common factor in intermediate filament inclusion bodies of diverse type in man, including those of Parkinson's disease, Pick's disease, and Alzheimer's disease, as well as Rosenthal fibres in cerebellar astrocytomas, cytoplasmic bodies in muscle, and mallory bodies in alcoholic liver disease. J Pathol 1988;155:9-15.
2. Alves-Rodrigues A, Gregori L, Figueiredo-Pereira ME. Ubiquitin, cellular inclusions and their role in neurodegeneration. Trends Neurosci 1998; 21:516-520.
3. Hoyt MA, Coffino P. Ubiquitin-free routes into the proteasome. Cell Mol Life Sci 2004; 61:1596-1600.
4. Prakash S, Tian L, Ratliff KS et al. An unstructured initiation site is required for efficient proteasome-mediated degradation. Nat Struct Mol Biol 2004; 11:830-837.
5. Beal R, Deveraux Q, Xia G et al. Surface hydrophobic residues of multiubiquitin chains essential for proteolytic targeting. Proc Natl Acad Sci USA 1996; 93:861-866.
6. Dubiel W, Gordon C. Ubiquitin pathway: Another link in the polyubiquitin chain? Curr Biol 1999; 9:R554-R557.
7. Ciechanover A, Ben Saadon R. N-terminal ubiquitination: More protein substrates join in. Trends Cell Biol 2004; 14:103-106.
8. Cadwell K, Coscoy L. Ubiquitination on nonlysine residues by a viral E3 ubiquitin ligase. Science 2005; 309:127-130.
9. Koegl M, Hoppe T, Schlenker S et al. A novel ubiquitination factor, E4, is involved in multiubiquitin chain assembly. Cell 1999; 96:635-644.
10. Schwartz AL, Ciechanover A. The ubiquitin-proteasome pathway and pathogenesis of human diseases. Annu Rev Med 1999; 50:57-74.
11. Pickart CM. Back to the future with ubiquitin. Cell 2004; 116:181-190.
12. Nalepa G, Wade HJ. Therapeutic anti-cancer targets upstream of the proteasome. Cancer Treat Rev 2003; 29(Suppl 1):49-57.
13. Hoppe T. Multiubiquitylation by E4 enzymes: 'One size' doesn't fit all. Trends Biochem Sci 2005; 30:183-187.
14. McDonough H, Patterson C. CHIP: A link between the chaperone and proteasome systems. Cell Stress Chaperones 2003; 8:303-308.
15. Imai Y, Soda M, Hatakeyama S et al. CHIP is associated with Parkin, a gene responsible for familial Parkinson's disease, and enhances its ubiquitin ligase activity. Mol Cell 2002; 10:55-67.
16. Chung CH, Baek SH. Deubiquitinating enzymes: Their diversity and emerging roles. Biochem Biophys Res Commun 1999; 266:633-640.
17. Wilkinson KD. Regulation of ubiquitin-dependent processes by deubiquitinating enzymes. FASEB J 1997; 11:1245-1256.
18. DeMartino GN, Slaughter CA. The proteasome, a novel protease regulated by multiple mechanisms. J Biol Chem 1999; 274:22123-22126.
19. Groll M, Ditzel L, Lowe J et al. Structure of 20S proteasome from yeast at 2.4 A resolution. Nature 1997; 386:463-471.
20. Orlowski M, Cardozo C, Michaud C. Evidence for the presence of five distinct proteolytic components in the pituitary multicatalytic proteinase complex. Properties of two components cleaving bonds on the carboxyl side of branched chain and small neutral amino acids. Biochemistry 1993; 32:1563-1572.
21. Boes B, Hengel H, Ruppert T et al. Interferon gamma stimulation modulates the proteolytic activity and cleavage site preference of 20S mouse proteasomes. J Exp Med 1994; 179:901-909.

22. Ramos PC, Hockendorff J, Johnson ES et al. Ump1p is required for proper maturation of the 20S proteasome and becomes its substrate upon completion of the assembly. Cell 1998; 92:489-499.
23. Glickman MH, Rubin DM, Coux O et al. A subcomplex of the proteasome regulatory particle required for ubiquitin-conjugate degradation and related to the COP9-signalosome and eIF3. Cell 1998; 94:615-623.
24. Braun BC, Glickman M, Kraft R et al. The base of the proteasome regulatory particle exhibits chaperone-like activity. Nat Cell Biol 1999; 1:221-226.
25. Scherrer K, Bey F. The prosomes (multicatalytic proteinases; proteasomes) and their relationship to the untranslated messenger ribonucleoproteins, the cytoskeleton, and cell differentiation. Prog Nucleic Acid Res Mol Biol 1994; 49:1-64.
26. Palmer A, Rivett AJ, Thomson S et al. Subpopulations of proteasomes in rat liver nuclei, microsomes and cytosol. Biochem J 1996; 316(Pt 2):401-407.
27. Reits EAJ, Benham AM, Plougastel B et al. Dynamics of proteasome distribution in living cells. EMBO J 1997; 16:6087-6094.
28. Enenkel C, Lehmann A, Kloetzel PM. GFP-labelling of 26S proteasomes in living yeast: Insight into proteasomal functions at the nuclear envelope/rough ER. Mol Biol Rep 1999; 26:131-135.
29. Plemper RK, Wolf DH. Retrograde protein translocation: ERADication of secretory proteins in health and disease. Trends Biochem Sci 1999; 24:266-270.
30. Tran PB, Miller RJ. Aggregates in neurodegenerative disease: Crowds and power? Trends Neurosci 1999; 22:194-197.
31. Mezey E, Dehejia A, Harta G et al. Alpha synuclein in neurodegenerative disorders: Murderer or accomplice? Nat Med 1998; 4:755-757.
32. Madura K. Rad23 and Rpn10: Perennial wallflowers join the melee. Trends Biochem Sci 2004; 29:637-640.
33. Hartmann-Petersen R, Semple CA, Ponting CP et al. UBA domain containing proteins in fission yeast. Int J Biochem Cell Biol 2003; 35:629-636.
34. Seibenhener ML, Babu JR, Geetha T et al. Sequestosome 1/p62 is a polyubiquitin chain binding protein involved in ubiquitin proteasome degradation. Mol Cell Biol 2004; 24:8055-8068.
35. Wilkinson CR, Seeger M, Hartmann-Petersen R et al. Proteins containing the UBA domain are able to bind to multi-ubiquitin chains. Nat Cell Biol 2001; 3:939-943.
36. Raasi S, Orlov I, Fleming KG et al. Binding of polyubiquitin chains to ubiquitin-associated (UBA) domains of HHR23A. J Mol Biol 2004; 341:1367-1379.
37. Park I, Chung J, Walsh CT et al. Phosphotyrosine-independent binding of a 62-kDa protein to the src homology 2 (SH2) domain of p56lck and its regulation by phosphorylation of Ser-59 in the lck unique N-terminal region. Proc Natl Acad Sci USA 1995; 92:12338-12342.
38. Kuusisto E, Salminen A, Alafuzoff I. Ubiquitin-binding protein p62 is present in neuronal and glial inclusions in human tauopathies and synucleinopathies. Neuroreport 2001; 12:2085-2090.
39. Zatloukal K, Stumptner C, Fuchsbichler A et al. p62 is a common component of cytoplasmic inclusions in protein aggregation diseases. Am J Pathol 2002; 160:255-263.
40. Furukawa Y, Iseki E, Hino H et al. Ubiquitin and ubiquitin-related proteins in neurons and dendrites of brains of atypical Pick's disease without Pick bodies. Neuropathology 2004; 24:38-45.
41. Nakano T, Nakaso K, Nakashima K et al. Expression of ubiquitin-binding protein p62 in ubiquitin-immunoreactive intraneuronal inclusions in amyotrophic lateral sclerosis with dementia: Analysis of five autopsy cases with broad clinicopathological spectrum. Acta Neuropathol (Berl) 2004; 107:359-364.
42. Stumptner C, Fuchsbichler A, Heid H et al. Mallory body—a disease-associated type of sequestosome. Hepatology 2002; 35:1053-1062.
43. Nagaoka U, Kim K, Jana NR et al. Increased expression of p62 in expanded polyglutamine-expressing cells and its association with polyglutamine inclusions. J Neurochem 2004; 91:57-68.
44. Kuusisto E, Suuronen T, Salminen A. Ubiquitin-binding protein p62 expression is induced during apoptosis and proteasomal inhibition in neuronal cells. Biochem Biophys Res Commun 2001; 280:223-228.
45. Nakaso K, Yoshimoto Y, Nakano T et al. Transcriptional activation of p62/A170/ZIP during the formation of the aggregates: Possible mechanisms and the role in Lewy body formation in Parkinson's disease. Brain Res 2004; 1012:42-51.
46. Wang Z, Figueiredo-Pereira ME. Inhibition of sequestosome 1/p62 up-regulation prevents aggregation of ubiquitinated proteins induced by prostaglandin J2 without reducing its neurotoxicity. Mol Cell Neurosci 2005; in press.
47. Shin J. P62 and the sequestosome, a novel mechanism for protein metabolism. Arch Pharm Res 1998; 21:629-633.

48. Doss-Pepe EW, Stenroos ES, Johnson WG et al. Ataxin-3 interactions with rad23 and valosin-containing protein and its associations with ubiquitin chains and the proteasome are consistent with a role in ubiquitin-mediated proteolysis. Mol Cell Biol 2003; 23:6469-6483.
49. Burnett BG, Pittman RN. The polyglutamine neurodegenerative protein ataxin 3 regulates aggresome formation. Proc Natl Acad Sci USA 2005; 102:4330-4335.
50. Riley BE, Xu Y, Zoghbi HY et al. The effects of the polyglutamine repeat protein ataxin-1 on the UbL-UBA protein A1Up. J Biol Chem 2004; 279:42290-42301.
51. Raasi S, Pickart CM. Rad23 ubiquitin-associated domains (UBA) inhibit 26 S proteasome-catalyzed proteolysis by sequestering lysine 48-linked polyubiquitin chains. J Biol Chem 2003; 278:8951-8959.
52. Johnston JA, Ward CL, Kopito RR. Aggresomes: A cellular response to misfolded proteins. J Cell Biol 1998; 143:1883-1898.
53. Kopito RR. Aggresomes, inclusion bodies and protein aggregation. Trends Cell Biol 2000; 10:524-530.
54. Garcia-Mata R, Bebok Z, Sorscher EJ et al. Characterization and dynamics of aggresome formation by a cytosolic GFP- chimera. J Cell Biol 1999; 146:1239-1254.
55. Johnston JA, Illing ME, Kopito RR. Cytoplasmic dynein/dynactin mediates the assembly of aggresomes. Cell Motil Cytoskeleton 2002; 53:26-38.
56. Mallik R, Gross SP. Molecular motors: Strategies to get along. Curr Biol 2004; 14:R971-R982.
57. Fabunmi RP, Wigley WC, Thomas PJ et al. Activity and regulation of the centrosome-associated proteasome. J Biol Chem 2000; 275:409-413.
58. Wigley WC, Fabunmi RP, Lee MG et al. Dynamic association of proteasomal machinery with the centrosome. J Cell Biol 1999; 145:481-490.
59. Lelouard H, Gatti E, Cappello F et al. Transient aggregation of ubiquitinated proteins during dendritic cell maturation. Nature 2002; 417:177-182.
60. Lelouard H, Ferrand V, Marguet D et al. Dendritic cell aggresome-like induced structures are dedicated areas for ubiquitination and storage of newly synthesized defective proteins. J Cell Biol 2004; 164:667-675.
61. DeFillipo AM, Dai J, Li Z. Heat shock-induced dendritic cell maturation is coupled by transient aggregation of ubiquitinated proteins independently of heat shock factor 1 or inducible heat shock protein 70. Mol Immunol 2004; 41:785-792.
62. Canadien V, Tan T, Zilber R et al. Cutting edge: Microbial products elicit formation of dendritic cell aggresome-like induced structures in macrophages. J Immunol 2005; 174:2471-2475.
63. Kopito RR, Sitia R. Aggresomes and Russell bodies. Symptoms of cellular indigestion? EMBO Rep 2000; 1:225-231.
64. Russell W. Address on a characteristic organism of cancer. Br Med J 1890; 2:1356-1360.
65. Valetti C, Grossi CE, Milstein C et al. Russell bodies: A general response of secretory cells to synthesis of a mutant immunoglobulin which can neither exit from, nor be degraded in, the endoplasmic reticulum. J Cell Biol 1991; 115:983-994.
66. Trushina E, Heldebrant MP, Perez-Terzic CM et al. Microtubule destabilization and nuclear entry are sequential steps leading to toxicity in Huntington's disease. Proc Natl Acad Sci USA 2003; 100:12171-12176.

Chapter 7

The Role of Heat Shock Proteins during Neurodegeneration in Alzheimer's, Parkinson's and Huntington's Disease

Andreas Wyttenbach* and André Patrick Arrigo

Abstract

A number of acute and chronic neurodegenerative conditions are associated by protein misfolding and aggregation of proteins within and outside cells. Misfolded proteins and protein aggregation are controlled by molecular chaperones such as heat shock proteins (HSPs) that are constitutively and inducibly expressed in the nervous system. There is increasing evidence that HSPs could counteract common pathological mechanisms that take place during Alzheimer's disease (AD), Parkinson's disease (PD) and Huntington's disease (HD). This is achieved by HSPs either interfering with the misfolded disease proteins preventing unwanted interactions with other cellular proteins and/or by reducing the risk of formation of toxic oligomeric assemblies of the respective disease proteins such as tau and amyloid-β in AD, α-synuclein in PD and huntingtin in HD. But HSPs are also expected to interfere with detrimental processes that occur during these diseases including oxidative stress and abnormal activation of signaling pathways or act supportive towards degradation systems such as the ubiquitin proteasome- and the autophagic-lysosomal pathway. Specific neuronal structures such as synapses and axons also harbour HSPs that may be misregulated during the disease process. Hence HSPs are expected to be critically involved in the progression of AD, PD and HD making them potential therapeutic targets and the studies discussed in this chapter support this view.

Introduction

The misfolding and progressive polymerisation of otherwise soluble proteins is a common characteristic of a variety of diseases associated with neurodegeneration. These diseases include Alzheimer's disease (AD), Parkinson's disease (PD), amyotrophic lateral sclerosis (ALS), the prion disorders, tauopathies (causing frontotemporal dementias such as progressive supranuclear palsy and corticobasal degeneration), other synucleinopathies than PD (multiple system atrophy, dementia with Lewy bodies, progressive autonomic failure, rapid-eye-movement sleep disorders) and the polyglutamine (polyQ) diseases. Neuropathology of these disorders is caused by environmental factors and genetic mutations. In this chapter we will discuss the role of heat shock proteins (HSPs) in some of these disorders with a focus on AD and PD which are the most common neurodegenerative disorders associated with cognitive and motor impairments

*Corresponding Author: Andreas Wyttenbach—Southampton Neuroscience Group, School of Biological Sciences, University of Southampton, Bassett Crescent East, Southampton SO16 7PX, U.K. Email: aw3@soton.ac.uk

Heat Shock Proteins in Neural Cells, edited by Christiane Richter-Landsberg.

and Huntington's disease (HD) which is the most common inherited disease caused by an expansion of a polyQ tract in the associated disease gene. The molecular and cellular details vary enormously among these diseases, but the tendency of soluble proteins to develop an altered conformation and aggregate inside and/or outside cells appears to precede early clinical signs in each disease. Protein misfolding and aggregation is associated with profound neuronal dysfunction and death, but the reader should note that currently it is still unclear whether this phenomenon explains the fundamental basis of these diseases or represents a crucial secondary step in the course of the disorder.[1]

Misfolded proteins are recognized by a set of conserved intracellular proteins known as molecular chaperones that mediate correct folding, assembly and degradation of proteins.[2] The heat shock proteins (HSPs) are a family of proteins that can act as molecular chaperones and are classified into six major families: Hsp100, Hsp90, Hsp70, Hsp60, Hsp40 and small HSPs (sHSPs, e.g., αB crystallin and Hsp27). Because HSPs are constitutively expressed and/or inducibly regulated in the nervous system and prevent aggregation of unfolded/misfolded polypeptides, assist refolding and play a role during solubilization of stable protein aggregates, they may be crucial modulators of neurotoxicity in AD, PD and HD. Here we are reviewing recent progress from many laboratories that support this view. After a brief introduction to each disorder we will discuss whether and how HSPs could impact on the common molecular and cellular mechanisms implicated in all three disease pathologies.

Alzheimer's Disease

AD pathology can occur in a sporadic manner (most cases) or is dominantly inherited (ca. 5% of all AD cases).[3] Symptoms are progressive and include difficulties in acquisition of new memory, memory loss and depression that typically occur with late onset of the disease. Idiopathic cases of AD appear to be phenocopies of inherited cases, but inherited AD onset occurs earlier in life. AD is the most prevalent human dementia affecting 10% of the population over 65.[3] AD has been defined by the occurrence of two types of lesions in brain regions serving memory and cognition: the neurofibrillary tangles (NFTs) and the amyloid plaques. NFTs consist of paired, helically wound or straight protein filaments made of the pathologically altered microtubule (MT)-associated protein tau (Fig. 1A).[4] Tau proteins stabilize and promote MT polymerization in neuronal perikarya and processes. NFTs are normally found in the cytoplasm and neuritic processes of affected nerve cells in selectively vulnerable regions of the central nervous system (CNS) in AD.[4] Hyperphosphorylation dislodges tau from the MT surface, potentially resulting in compromised axonal integrity and toxic tau peptides. Recent biochemical and animal model studies indicate that phosphorylation of tau's MT-binding domain by MARK (MAP/microtubule affinity regulating kinase) may prime tau for hyperphosphorylation by kinases such as GSK-3 and Cdk5 that in turn trigger the aggregation of tau.[5] Filamentous tau tangles are made of hyperphosphorylated and otherwise abnormal tau. The filaments vary between 8-20nm in width and it is the repeat region of tau that forms the core (see Fig. 1). Filaments show a clear cross-β structure, the defining feature of amyloid fibres. A likely mechanism for tau fibrillization is that the protein disengages from MTs and hence the pool of soluble tau increases. Soluble tau may be less degradable and more prone to aggregation, likely due to abnormal phosphorylation events. The transition of the largely unstructured, normal tau into a β-sheet containing species may be facilitated by intracellular interactions with membranes and organelles. Nucleation of tau oligomers and filaments and the formation of mature NFTs likely ocurs via a nonfibrillar step. Amorphous deposits of tau have indeed been detected in human brain and may represent such a "pretangle" stage. Although tau is predominantly a neuronal protein, tau pathologies occur in oligodendrocytes and astrocytes in diverse neurodegenerative disorders.[6] For example, the identification of tau mutations in families with frontotemporal dementia with parkinsonism linked to chromosome 17 (FTDP-17) show that mutations reduce binding to MTs and increase NFT formation.[6]

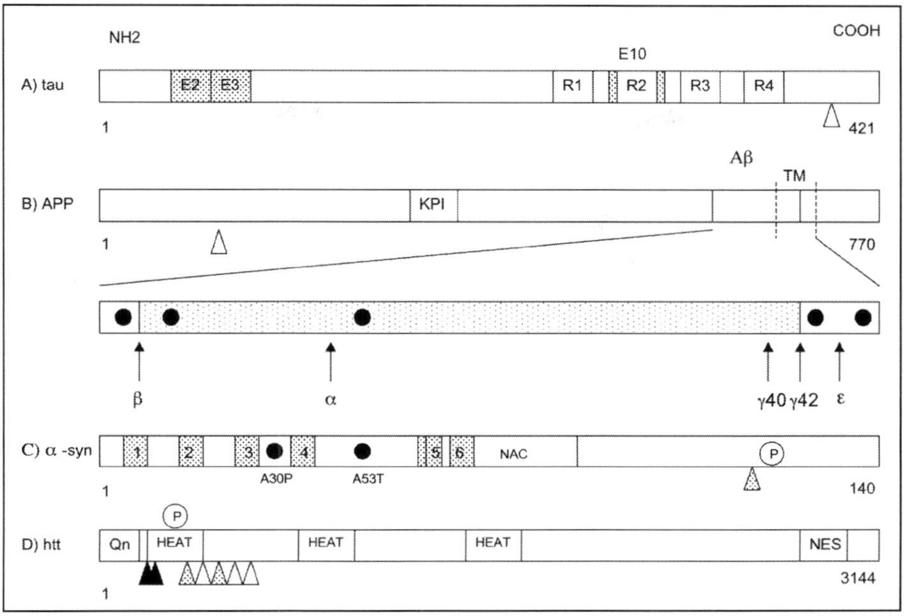

Figure 1. A) In the adult human brain six tau isoforms are produced as a result of alternative mRNA splicing of E2, E3 and E10 (shaded areas). The black bars represent the 18 amino acid microtubule binding repeats (R1 to R4). The isoforms differ by the presence of three or four C-terminal tandem repeat sequences. There are 79 potential serine and threonine phosphate acceptor residues in the longest isoform of tau that is depicted here, and phosphorylation at ca. 30 of these, that are located close to the MT binding sites, have been reported in normal tau. Tau is cleaved at Asp 421 by multiple caspases which appear to increase tau aggregation (empty triangle). B) Schematic diagram of the largest splice form of APP (first line in diagram). APP is undergoing cleavages by metalloproteases (α-secretases) believed to be ADAM10 and ADAM 17. Cleavage by α-secretase (α, second line in diagram) occurs mainly between residue 16 and 17 in the Aβ region of APP and hence cannot generate Aβ peptides, but enables secretion of a soluble ectodomain into e.g., the extracellular space. Sequential cleavage of APP by two other proteolytic complexes namely β-secretase (also called BACE, β) and γ-secretase (γ), an intramembrane aspartyl protease with presenilin at its catalytic site, results in the production of the Aβ peptide (shaded area of second line). Aβ includes 28 residues just outside the membrane and the first 12-14 residues of the transmembrane domain (TM). Cleavage at the epsilon (ε) site releases the APP intracellular domain (AICD) into the cytoplasm. There are other processing events of APP that are relevant for several biological activities that are not considered here (for review see ref 1). Disease causing mutations in APP are indicated by filled circles at the β/γ site in second line. Unfilled triangles indicate a caspase cleavage site within APP; KPI: serine protease inhibitor domain of the Kunitz type. C) The synuclein proteins (α, β, γ) are composed of a N-terminal region with 5-6 imperfectly conserved repeats (KTKEGV, shaded area number 1-6). The conserved N-terminal repeat domain of α-synuclein is thought to mediate lipid binding and dimerisation. α-synuclein can be ubiquitinated, glycosylated, nitrated and phosphorylated at Serine 129 (empty circle containing P) that may affect its aggregation. Additionally, a calpain cleavage site at Asn122 has been identified (shaded triangle) that could affect α-synuclein stability and aggregation. Such a truncated version of α-synuclein has also been identified in vivo. The A30P and A53T mutations are indicated by filled circles. D) Huntingtin protein. The protein is cleaved by various caspases (unfilled triangles), calpains (shaded triangles) or unknown proteases (filled triangles). Htt is phosphorylated by protein kinase B (PKB/Akt), the serum and glucocorticoid induced kinase (GSK) and Cdk5. Htt contains HEAT-motifs that are protein interaction domains. The polyQ expansion is located in Exon 1 (Qn, box). NES: Nuclear export signal.

Amyloid plaques are the other major lesion in AD and are extracellular deposits of 8-10 nm wide amyloid fibrils that result from amyloid-β (Aβ) protein polymerisation.[1] The prevalent form of amyloid plaques are known as the 'neuritic plaque' (NPs). NPs comprise an extracellular core of amyloid fibrils encircled by dystrophic dendrites and axons that can also contain paired helical filaments (PHFs) of tau, activated microglia and reactive astrocytes. [1] NPs are found in the molecular layer of the dentate gyrus of the hippocampus, the amygdala, the association cortices of the frontal, temporal and parietal lobes and also in certain deep brain nuclei projecting to these areas. There are other morphologically distinct forms of Aβ plaques such as "diffuse plaques" found in the limbic and association cortices that are characteristically loose/granular and lacking dystrophic neurites. [1] Aβ is a 38-43 residue proteolytic fragment of the amyloid precursor protein (APP). APP is a ubiquitously expressed type 1 membrane glycoprotein that undergoes complex processing and is produced normally throughout life by most mammalian cells. Some of the elaborate processing steps of APP are explained in Figure 1B. The genes implicated in familial AD to date are presenilin-1 and presenilin-2 (PS1/2), APP and the ε4 allele of apolipoprotein E (ApoE4).[1] Disease causing mutations in APP, PS1/2 and ApoE4 inheritance result in either increased Aβ42 production/ oligomerisation or reduced clearance.

The prevailing view that extracellular Aβ deposition is the primary toxic insult in AD has been challenged because a number of studies suggest that intracellular Aβ accumulation may play an early and relevant pathological role.[7] Aβ42 appears to be the prominent form accumulating inside neurons in the brains of AD patients and transgenic mice expressing familial AD mutations.[7]

The sequence of events underlying the development of AD and the role of the assemblies of tau and/or Aβ is not known. The genetics of AD suggest that Aβ accumulation is the factor that could trigger NFT formation. Based on available evidence it seems that APP molecules at the plasma membrane and in intracellular vesicles are cleaved to liberate the Aβ region (Figs. 1B,2). A portion of Aβ peptides may then oligomerize, initially intravesicularly, and then be released into the interstitial fluid of the brain resulting in interference with synaptic function.[1] Aβ can further polymerize into insoluble fibrils that form plaques that in turn may cause dysfunction of axons and dendrites. Concomitant intraneuronal cytoplasmic kinase activation may then lead to hyperphosphorylation of tau, its dislocation from MTs and aggregation into NFTs resulting in neuritic damage (Fig. 2). The reader should note that not everybody agrees with this view. Activated microglia and reactive astrocytes surrounding the plaques also contribute to neuropathology that may be causally connected to oxidative stress, abnormal activation of signalling pathways, the formation of amyloid ion channels and altered degradation systems (see section *Common Mechanisms Leading to Impaired Neuronal Function and Death and the Role of HSPs*). It is the diffusable, soluble oligomers of Aβ that are hypothesised to directly compromise neuronal functions first rather than fibrils and plaques.[1] Interestingly, recent evidence suggest a connection in AD and PD pathology where tau and α-synuclein inclusions are found in the same cell and a direct link between aggregation of tau and α-synuclein has been demonstrated.[6]

Parkinson's Disease

Clinical features of PD include tremor, rigidity and bradykinesia (slowness of voluntary movements) and are found in 0.2% of the population.[8] The majority of the cases seem to occur sporadically, but in rare cases the disease can also be inherited in an autosomal dominant or recessive fashion. In 1997, a point mutation (A53T) in a gene encoded by α-synuclein was identified in an Italian family and subsequently a second PD-linked dominant mutation in α-synuclein (A30P) was found in a German family.[9] Meanwhile other mutations linked to PD in genes encoding parkin, ubiquitin C-terminal hydrolase -L1 (UCH-L1), DJ-1 and PTEN-induced kinase 1 (Pink-1) have been identified.[9] As in AD, the infrequent familial forms of PD are expected to be highly informative for the understanding of the pathogenesis of

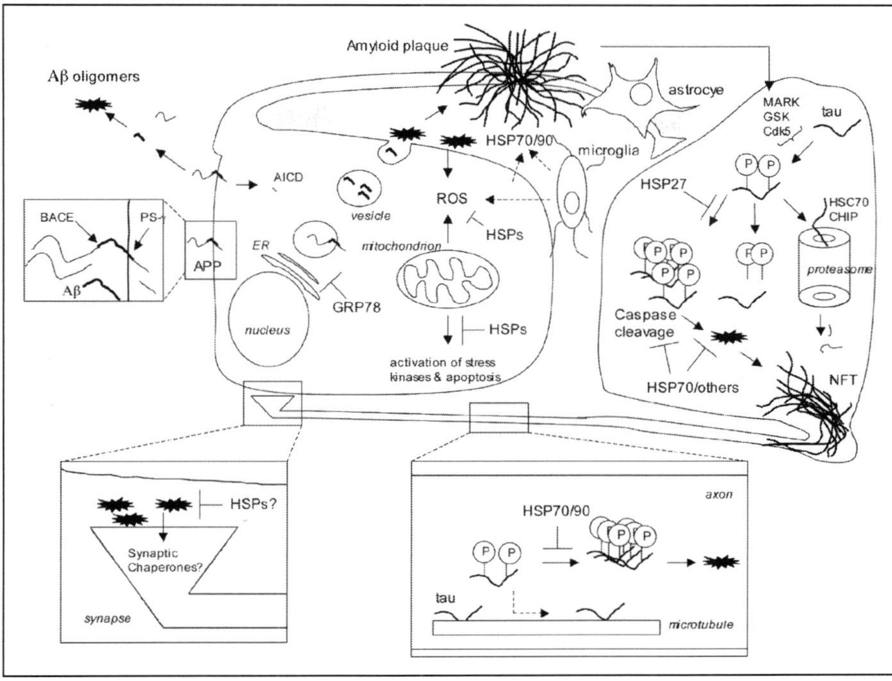

Figure 2. Several mechanisms have been proposed to contribute to AD progression. Two neurons are shown with APP -> Aβ pathology on the left and tau pathology on the right of the scheme. HSP90 accumulates in extracellular plaques and it has been suggested that HSPs outside neurons play a role in the phagocytic digestion of amyloid plaques by microglia. Non-neuronal cells may contribute to the increased oxidative damage found in AD brains which could be counteracted by HSPs. GRP78 (an ER HSP70 member) can bind to APP and decrease secretion of Aβ40/42. Direct interaction with HSPs inside or outside cells are likely to regulate Aβ assembly, e.g., prevent pore formation (not shown in the scheme). HSPs are expected to modulate the production of reactive oxygen species (e.g., HSP27). It remains to be seen if HSPs modulate the activation of stress kinases (e.g., Jun N-terminal kinase (JNK), see text) implicated in AD or interfere with the various steps in the apoptotic cascade (also see the scheme for HD in Fig. 4). At the synapse intracellular and extracellular HSPs may also prevent toxic interactions of Aβ with the membrane. Specific synaptic chaperones may be directly impaired by toxic Aβ assemblies. HSP27 interacts preferentially with hyperphosphorylated tau and by keeping its levels low, increases dephosphorylated tau hence suppressing toxicity. HSP70 prevents tau aggregation and may also interfere with caspases that could accelerate tau aggregation through cleavage. HSP70/HSP90 do functionally interact with tau in vitro and prevent the accumulation of insoluble tau and thereby may restore impaired axonal transport. Finally, CHIP and HSP70 facilitate tau ubiquitylation and suppress toxicity.

the idiopathic form. The dopaminergic neurons in the substantia nigra (SN) and the noradrenergic neurons in the locus coeruleus frequently exhibit characteristic filamentous inclusion bodies (IBs) in PD called "Lewy bodies" (LBs). There is a depletion of dopamine in the striatum and substantial nerve cell loss in the SN during the disease.[9] The main component of LBs is wild type α-synuclein.[8] α-synuclein is a 'natively unfolded' normally soluble protein which under conditions of oxidative stress (including mitochondrial complex I deficiency) forms amyloid-like filaments. Under physiological conditions in vitro, α-synuclein lacks a well-defined structure and hence is an intrinsically unstructured protein, similar to tau. α-Synuclein is a 140 amino acid protein that is predominantly expressed in CNS neurons, where it is concentrated at presynaptic terminals. The function of α-synuclein is not fully characterised but various studies suggest that it is involved in modulating synaptic transmission, regulating the localisation of vesicles in synapses and participating in neuronal plasticity. Human α-synuclein was identified as the precursor protein for the nonamyloid beta component (NAC) of AD plaques. The hydrophobic NAC sequence is important for fibrillization (see Fig. 1C). Like the plaques and tangles in AD, LBs may not initiate the degenerative process as a comparison of neurons in the SN with and without LBs failed to show any obvious differences including apoptotic-like changes. Furthermore, mutations in the parkin gene, linked to PD, that appears to cause degeneration of neurons in the SN, do not result in LB formation (see below). Also the lack of a correlation between the PD causing mutation in α-synuclein and the acceleration of fibril formation suggest that the fibril is not the pathogenic entity. These and other observations have shifted the attention of researchers towards the role of α-synuclein protofibrils which are transient β-sheet containing oligomers. These species have been proposed to form porelike structures that may potentially disrupt ionic and/or metabolic homeostasis.[8]

Mutations in parkin are the most common cause of autosomal recessive early-onset parkinsonism.[9] Parkin mutations have been found in about 50% of familial cases of PD and in 10-20% of cases without a positive family history. Parkin is a protein exhibiting ubiquitin ligase activity.[9] The clinical features of parkin-related diseases can be similar to those of late onset sporadic PD (see above). Mutations in parkin result in a loss-of-function of E3 and substrates of parkin abnormally accumulate within dopaminergic neurons. LB formation generally does not occur in parkin-related diseases and there is severe neuronal loss and gliosis in brains of these patients. Several proteins implicated in oxidative stress pathways are down-regulated in parkin deficient mice suggesting that parkin functions are related to oxidative stress control.[9] This finding fits with the observation that idiopathic PD may be triggered by chronic exposure to environmental toxins such as rotenone, paraquat or MPTP, all of which increase free radical production and induce LB-like inclusions containing α-synuclein in dopaminergic neurons (Fig. 3). The third gene implicated in PD is UCH-L1 that is an abundant de-ubiquitilating enzyme and hence can be placed in a pathway potentially related to parkin.[9] DJ-1 is a cytoplasmic neuronal protein with chaperone activity induced under conditions of oxidative stress and Pink-1 is a mitochondrial protein kinase.[9] For the moment it is unclear if these genes are targeting similar cellular and molecular mechanisms implicated in PD. However, there is strong evidence that oxidative stress, mitochondrial dysfunction and impairment of the ubiquitin-proteasome pathway (UPP) is central to the neuropathology of PD (Fig. 3).

Huntington's Disease

HD is a fatal neurodegenerative disorder and together with spinobulbar muscular atrophy (SBMA), dentatorubral pallidoluysian atrophy (DRPLA), and spinocerebellar ataxia (SCA) types, 1, 2, 3, 6, 7, and 17 known as a polyglutamine (polyQ) disease.[10] All the polyQ diseases are inherited in a strictly autosomal dominant manner with the exception of SBMA that is X-linked. These diseases are caused by a polyQ expansion mutation. Mutant proteins between diseases do not share any homology except the polyQ tract: this indicates common molecular mechanisms of pathology during polyQ pathology. HD is characterized by abnormal movements including chorea, dementia and emotional disorders and affects ca. one individual out of ten thousand and hence is considered a rare disorder. Age of onset of polyQ diseases correlates inversely with repeat number of the expansion and HD is no exception: longer polyQ repeats lead to earlier age of onset and a more severe pathology.[10] The pathologic length of the polyQ repeat varies for each polyQ disease, but is generally $\approx$ 40 or greater. At the HD locus unaffected individuals typically show 11-34 glutamine residues while in HD patients the polyQ stretch expands to >36 glutamines located in Exon 1 of the HD gene that consists of 67 Exons (see Fig. 1D).

Huntingtin (htt) is a large protein comprising 3144 amino acids containing a functional nuclear export signal (NES) at its C-terminus. Full-length htt shuttles between the cytoplasm

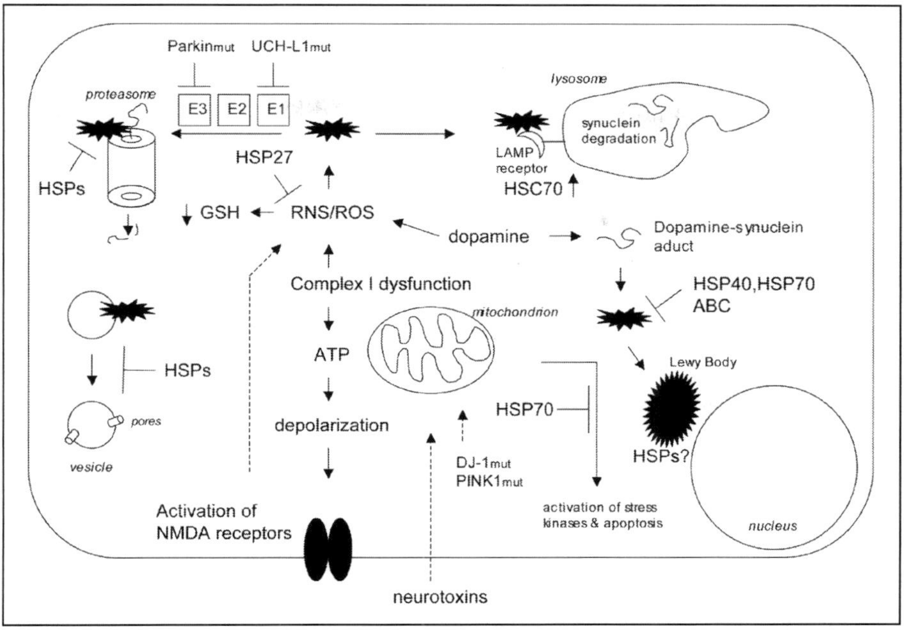

Figure 3. Current evidence suggests that complex I inhibition may be the central cause of sporadic PD and that this defect causes aggregation of α-synuclein which then contributes to the demise of dopamine neurons. Several environmental toxins (e.g., herbicides, insecticides such as rotenone) are selectively neurotoxic to dopaminergic neurons and inhibit complex I of the electron transfer chain within mitochondria. These toxins also lead to the formation of LBs. Several known neuronal forms of death are activated by complex I derangement including apoptosis that can be inhibited by HSPs. ROS/RNS play a critical role during PD and hence HSPs that impact on the redox status of the cell (e.g., HSP27) are expected to modulate PD toxicity. Impairment of the UPS may contribute to the demise of dopamine neurons through direct proteasome impairment by α-synuclein assemblies or indirectly via mutations in parkin and UCH-L1 (see text). HSP70 interacts with a-synuclein oligomers and with HSP40 may decrease aggregation and/or promote degradation via the proteasome. HSPs may facilitate chaperone-mediated autophagy (CMA) (α-synuclein is degraded by CMA), especially HSC70 that regulates the internalisation of protein via LAMP (lysosome-associated membrane protein). ABC appears to interact with α-synuclein and affect amyloid fibril formation. α-synuclein interacts with lipids and α-synuclein protofibrils may permeabilize vesicles. HSP70 binds preferentially to prefibrillar α-synuclein structures and hence may reduce the risk of permeabilisation. Abundant HSPs are found to localise with LBs and hence may affect its formation/degradation.

and the nucleus. If the NES is lost due to proteolytic cleavage by caspases (see Fig. 1D) and calpains, smaller fragments are formed that are more toxic and more likely to accumulate in the nucleus giving rise to IBs. Hence htt like APP, tau and synuclein is proteolytically regulated. Htt is associated with various organelles including the endoplasmatic reticulum, Golgi complex, nucleus, synaptic vesicles and mitochondria. Based on the known functions of the many htt interacting proteins it is believed that htt plays a role as a scaffolding protein orchestrating sets of proteins for signalling processes and intracellular transport. Wildtype htt may also have anti-apoptotic properties.

HD typically begins in midlife (but there is juvenile onset) and is characterised by progressive neuronal dysfunction and neuronal loss, and patients die 10 to 15 years after the onset of symptoms. The selective neurodegeneration in the brain occurs most prominently in the striatum and deep layers of the cerebral cortex.[10] In advanced stages of the disease other brain regions such as the hippocampus, hypothalamus, cerebellum and amygdala can also be affected. Projection neurons sending their axons to various brain regions are most severely affected. In the striatum the GABAergic medium-sized spiny neurons (MSNs) are most vulnerable. This phenomenon is observed despite ubiquitous expression of the HD gene. PolyQ expansions compromise neuronal function through several deleterious mechanisms which may vary during the long period of HD progression: transcriptional alterations, impairment of degradation systems such as the UPP, abnormal axonal transport and axonal degeneration, bio-energetic defects due to abnormal mitochondrial function and oxidative stress have all been implicated in HD.

The discovery of intraneuronal inclusions that are of a fibrillar nature and contain the mutant proteins in polyQ diseases and transgenic mice with polyQ mutations has provided a common denominator for pathology.[11] The polyQ expansion at the N-terminus of htt produces an abnormal conformation inducing a toxic gain-of-function in the mutant protein, a process that is associated with nucleation and subsequent aggregation. It is thought that polyQ fibrillogenesis is an ordered polymerization process similar to amyloid fibrillogenesis in AD and PD. The rate limiting step in the aggregation process is the formation of an oligomeric nucleus which can form after a repeat-length dependent conformational change of the polyQ monomer from a random coil to a parallel, helical β-sheet.[12] PolyQ aggregates and IBs are predominantly found in the nucleus and axonal processes, but cytoplasmic aggregates have also been described in the HD brain. Similar to the situation in AD and PD the oligomeric precursors of htt polyQ fibrils are thought to be the most neurotoxic. Htt aggregates have been shown to recruit a number of vital cellular proteins including transcription factors, HSPs and components of the UPP.

Common Mechanisms Leading to Impaired Neuronal Function and Death and the Role of HSPs

Direct Interference of HSPs with Misfolding, Oligomerisation and Fibrillogenesis of Disease Proteins

The assembly of misfolded proteins into amyloid fibrils and subsequently into IBs, plaques, NFTs or LBs is a complex process and could be similar among the various proteins that produce these structures. Research into the amyloid aggregation process starting from a partially folded peptide or the native protein to misfolded or unfolded monomeric α-synuclein, Aβ, tau and htt has revealed the production of several, distinct structures such as spherical or annular oligomers and amorphous aggregates and/or fibrils (Fig. 5). HSPs may interfere at various stages during this process by associating with misfolded disease proteins and higher order structures such as oligomers and IBs. This may result in the prevention of a toxic intra-molecular conformational change of the peptide itself or decrease the probability of unwanted homo-and/or heterotypic interactions thereby modifying amyloid assembly. Indeed, various HSPs have been shown to bind in vitro and/or in vivo to mutant htt[13], hyperphosphorylated tau[14] and α-synuclein prefibrillar structures[15] and modulate amyloid assembly. Consistent with this view, Schaffar and colleagues using fluorescence resonance energy transfer (FRET) showed that HSP40 and HSP70 prevent an intra-molecular conformational change in mutant htt.[16] That study also showed that HSPs prevent heterotypic interactions between polyQ expanded htt and transcription factors. Dou and colleagues reported an inverse relationship between aggregated tau and the level of HSP70/90 in transgenic mice and AD brain.[17] While increased levels of HSPs promoted tau solubility and binding to MTs, but reduced tau phosphorylation, lowering the levels of HSPs by use of RNA-mediated interference (RNAi) had the opposite effects.[17] Several in vitro and in vivo model studies of each disease show that artificial elevation of certain HSPs by over-expression decreases the amount of insoluble disease protein and/or deposits which is associated with a decrease in

Table 1. Several HSPs co-localise with protein deposits formed by tau, Aβ, huntingtin and α-synuclein in the human brain of affected individuals and modulate oligomerization and fibril formation in test tube experiments and in cell models of the respective diseases

Protein/ Peptide	Co-Localisation with Protein Deposits in Human Brain	Modulation of Aggregation Process in Vitro and/or in Vivo	Increase/Decrease in Affected Brain Regions
tau	HSP27	HSP27	
	HSP90	HSP70, HSP90	HSP70, HSP90
APP		GRP78, HSP90	
Aβ	HSP28	ABC	ABC, HSP27
	HSP27, ABC		
	HSP72		
Huntingtin	?	HSP22, HSP40, HSP70, HSP84	?
Synuclein	HSP40, HSP70 ABC, HSP27, HSP110	ABC, HSP40, HSP70	HSP27, ABC

HSP expression has also been reported to be up-or down-regulated in some of the brain regions most affected during the disease process (see text for details). Data are adapted from reference 22 and references therein.

toxicity in most of the cases (see Table 1). HSPs may reduce aggregation by interfering with the APP secretory pathway (Fig. 2): GRP78, an HSP70 localised in the ER, binds to APP decreasing the secretion of amyloid Aβ40 and Aβ42 in cell culture models.[18] Importantly, interactions of HSPs with one disease protein may affect the assembly process of another aggregation prone molecule: α-synuclein binds to tau and increases tau fibrillization and Aβ affects α-synuclein amyloid assembly.[6] α-synuclein may also increase htt aggregation and is localised to htt IBs in human brain.[19,20]

Some HSPs colocalise with intracellular NFTs and Aβ plaques in the extracellular space and in PD many HSPs are sequestered into LBs (see Table 1). Although the colocalisation of HSPs and IBs have not been carefully studied in the HD brain, there are several studies that showed a redistribution of HSP40/HSP70 into cytoplasmic and nuclear inclusions in the related polyQ disorders spinocerebellar ataxias (SCA) 1, 3 and 7.[13] It is unclear if such a redistribution reflects an irreversible sequestration and loss of function of HSPs or a failed attempt to refold aggregated proteins or both. A recent study addressed this question using florescence recovery after photobleaching (FRAP) and fluorescence loss in photobleaching (FLIP): the interaction between HSP70 and htt IBs exhibit rapid association and dissociation kinetics and hence HSP70 may actively regulate htt IB formation.[21]

A study on the assembly behaviour of a fragment of htt showed that similar to α-synuclein and Aβ, mutant htt assembles into spherical, annular and amorphous structures that might be on a pathway to fibril assembly.[22] HSP40 and HSP70 attenuated the formation of htt spherical and annular oligomers promoting the accumulation of fibrillar and amorphous aggregates that may be less toxic. This study proposed that HSPs could facilitate the folding of a specific misfolded monomeric conformation that allows on-pathway assembly to occur by monomeric addition to fibril nuclei, while decreasing the likelihood of interactions that promote the formation of off-pathway events during fibril formation (formation of spherical and annular oligomers). This model remains to be tested for other aggregation prone molecules (Fig. 5).

Prevention of Oxidative Stress by Heat Shock Proteins

The cause of neuronal dysfunction and death in each disease is multi-factorial, but evidence is emerging that the pathological depositions of abnormal proteins is associated with alterations in redox state homeostasis and mitochondrial dysfunction.[23] PD is characterized by increased oxidative damages, altered activities of antioxidant defence enzymes and decreased mitochondrial complex I activity (Fig. 3). Increased oxidative damage and altered mitochondrial activity (probably because of decreased complex IV activity) has also been reported in AD.[23] In the HD brain, oxidative damage has been detected together with a decreased mitochondrial complex II and IV activity[24] and in HD mouse models free radical damage has been observed.[25] Similar observations were made in a cellular model of HD.[26]

Evidence for the presence of oxidative stress can also be provided by the finding that (i) the pathological protein deposits can be recognized by antibodies that are specific to protein side-chains modified either directly by reactive oxygen or nitrogen species, or by products of lipid peroxidation or glycoxidation and (ii) the brain regions that are the most affected often contain deposits of redox-active transition metals, such as iron and copper. Moreover, membrane-associated oxidative stress (MAOS) which is a metal-catalyzed oxidative disruption of membrane protein and lipid signalling has been suggested to occur in the pathogenesis of AD, PD and HD.[27] The origin of the oxidative stress that is concomitant to the presence of aggregated neuronal proteins is not clear. In this respect, cellular models based on the expression of mutant htt have been useful to show that the oxidative events are probably not cell type specific since they can still be detected if the abnormal protein is expressed in cells other than neurones.[26] Moreover, the degree of aggregation of mutant htt, which is a direct consequence of the number of polyQ expansions, appears to be a key factor generating oxidative stress.[26] In vitro assays have been performed in which aggregated α-synuclein and β-amyloid are incubated in the presence of redox-active transition metals.[28] These assays revealed that iron and copper catalyze the formation of reactive oxygen species (ROS) by the aggregating proteins probably through Fenton-Haber-Weiss types of reaction. This suggests that an aggregating protein in contact with redox-active transition metals is capable of ROS generation. Htt is an iron-regulated protein and similar results could be found. Moreover, using a cellular model of HD[26], we recently noticed that the transient expression of mutant htt increased the intracellular level of iron, a phenomenon that correlated with the number of polyQ repeats (W. Firdaus, A. Wyttenbach, B. Currie, A.P. Arrigo, unpublished). It is not known if specific properties of some neuronal proteins are required for production of ROS in the presence of metals or if this phenomenon is more general and can be observed in proteins that are prone to aggregation.

It is well known that HSPs can protect cells against the damage induced by oxidative stress (Figs. 2-4). HSP mediated resistance against oxidative stress has been recently shown in an HD cellular model.[26] HSP mediated protection of cells against oxidative stress appears to occur via the molecular chaperone activity of these proteins. For example, in addition of being able to take care of the elimination of oxidized proteins, HSPs can protect specific enzymes involved in the detoxification of reactive oxygen and are involved in the reduction of oxidized glutathione.[29] It is also notable that the forced expression of Hsp70/Hsp40 and/or Hsp27 attenuated the intracellular increase in iron found in cells transiently expressing mutant htt polypeptides (W. Firdaus, A. Wyttenbach, B. Currie, A.P. Arrigo, unpublished). This finding is consistent with our observation that HSP27 plays a role in intracellular iron homeostasis.[29]

What is the importance of oxidative stress in neurodegenerative diseases? Delineation of the profile of oxidative damage in each disease must be determined as it will provide clues to how the specific neuronal populations are affected by the individual disease conditions. In this respect, studying single markers of oxidative damage outside the context of oxidative balance is probably not sufficient. Concerning anti-oxidants, in the case of a cellular model of HD it is clear that they abolish some of the intracellular cytotoxicity of the aggregates[26]

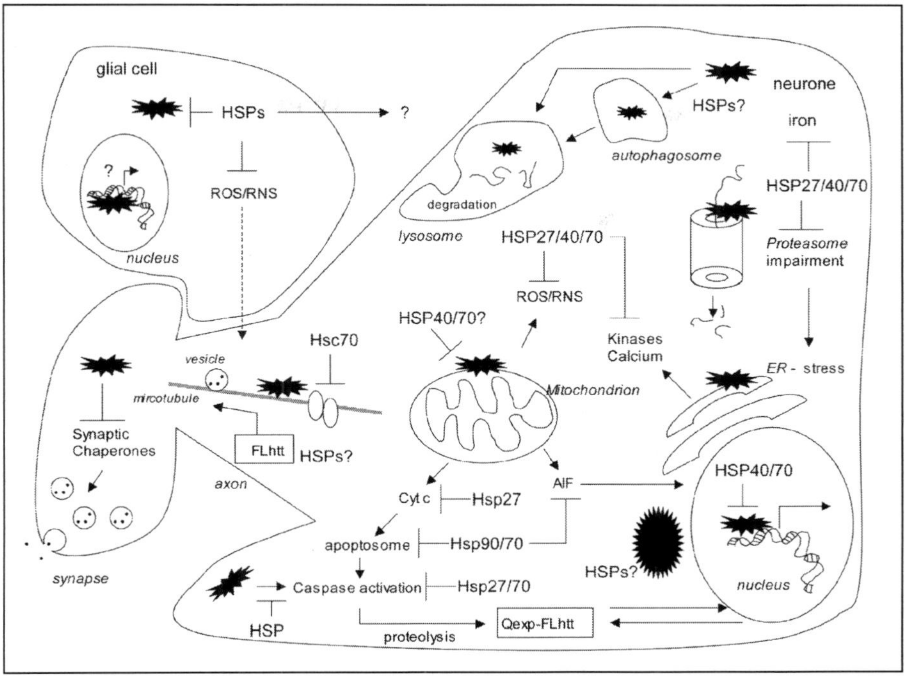

Figure 4. Mechanisms that have been implicated in HD neuropathology. It is likely that impairment of axonal transport, the synapse and transcription are early events followed by energetic/mitochondrial abnormalities, oxidative stress and alterations of degradation systems linked to the ageing process all lead to neuronal loss. HSPs act on these mechanisms as shown in this figure. Question marks indicate points of predicted interference by HSPs on htt toxicity that have not yet been demonstrated. The "explosion" (filled) represents polyQ expanded full-length htt (huntingtin) or a fragment of htt in a toxic, monomeric conformation or an early aggregation intermediate. Impaired axonal transport can arise from blockage by aggregates or impaired microtubule (MT) dependent vesicle transport that appears to be one of the normal functions of htt. Both could be counteracted by HSPs. Direct alterations of synaptic chaperones such as CSP (cystein string protein), SGTs (small glutamine-rich tetratricopeptide containing proteins) and HSC70 by an interaction with htt and/or depletion by IBs may contribute to synaptic degeneration. HSPs may affect abnormal interactions between polyQ stretches and transcription factors in the nucleus. Interactions of htt with membrane proteins, potential pore formation at e.g., the mitochondria and abnormal free radical generation could also be prevented by HSPs. HSPs are likely to inhibit caspases and death effectors released by mitochondria (cyt c: cytochrome c; AIF: apoptosis inducing factor). Proteasome impairment, ER-stress and associated UPR (unfolded protein response) may occur in HD and may be counteracted by HSPs. Activation of stress kinases can be negatively regulated by HSPs. The 32-star (left of the nucleus) represents an aggresome. Inclusion bodies found in the cytoplasm of HD neurones resemble aggresomes. HSPs associate with aggresomes and may regulate their function. Glial cells (microglia, astrocytes) are likely to contribute to neuronal pathology through their interactions with neurons at the synapse, axons and dendrites and as producers of free radicals. HSPs are highly inducible in glia compared to neural cells.

and various antioxidants are known to protect against Aβ toxicity in cell models. However, aggregates are still present hence suggesting that anti-oxidants will have only a transient beneficial effect. In vivo, testing of anti-oxidants is more problematic, however, because of the proximal role that transition metals appear to play in the pathology of neurodegenerative diseases.

Maintenance of Signalling Homeostasis: Can Heat Shock Proteins Suppress the Activation of Detrimental Signalling Pathways?

Genomic and proteomic screens on either diseased human tissue or cell and animal models accelerate the identification of signalling pathways that are impaired and/or abnormally activated in each disease. Many candidate pathways have not yet been functionally validated in appropriate disease models and although various pathways are modulated in a disease-specific manner, there are several signalling modules that are similarly activated in AD, PD and HD. Apoptotic pathways that typically involve caspase activation have been reported to be engaged in all three diseases. Evidence in PD suggests that a p53-glyceraldehyde-3 phosphate dehydrogenase (GAPDH) and/or FAS receptor-FADD-caspase 8 pathway converge onto Bax[30], a key neuronal death molecule that regulates the neuronal mitochondrial death pathway. Aβ production in both FAD and sporadic AD has been linked to the intrinsic, mitochondrial apoptotic pathway involving caspase-2, caspase-9 and Bax and the extrinsic receptor mediated apoptotic pathway (caspase-8) both converging onto the effector caspase-3.[31] Evidence for engagement of the mitochondrial pathway, caspase-1, -3, -8 and 9 activation and hence apoptosis in HD comes from studies on human brain tissue and many in vitro and in vivo models of the disease, but some evidence suggests that neuronal death in HD is not apoptotic and does not depend on Bax.[32] HSPs are known to interact with and suppress the activation of several molecules/arms of the apoptotic cascades (Figs. 2-4).[33] For example, HSP27, HSP70 and HSP90 all interfere with the formation of a functional apoptosome and are able to suppress caspase-9 activation (Fig. 4). HSP27 binds to cytochrome c and caspase-3 thereby negatively regulating cell death.[33] Therefore, HSPs are potentially important suppressors of death signalling pathways (Figs. 2-4). However, it is unlikely that only apoptotic pathways regulate neuronal death in these diseases. It will be crucial to elucidate the role of death pathways that involve the autophagic/lysosomal system and pathways linked to axonal degeneration upon which HSPs may act.

ER-stress sometimes associated with the unfolded protein response (UPR) has been associated with FAD and sporadic AD, PD and HD.[34] Common downstream effectors such as the ER stress kinases IRE and PERK are activated in these situations likely due to an overload of misfolded proteins. Here up-regulation of ER-resident chaperones of the HSP70 family such as GRP78/BiP normally occurs. There is indeed evidence of involvement of GRP78 in HD and AD.[18,35] A consequence of ER-stress in HD is the activation of Ask1, an upstream kinase that regulates pro-apoptotic JNK and p38MAPK activity (see below). Ask has also been implicated in Aβ induced neuronal death and since Ask1 is negatively regulated by HSP70 it could inhibit ER-stress activated kinases.

Stress activated mitogen-activated protein kinase (MAPKs) pathways appear to be modulated in AD, PD and HD and represent attractive therapeutic targets.[36] While there is strong evidence for over-activation of the p38 MAPK pathway in early stages of AD[36] the situation in PD and HD is less clear. Involvement of a JNK-dependent death pathway has been suggested in AD, PD and HD.[13,36] HSP70 inhibits JNK activity by either a direct interaction or by negatively regulating a JNK phosphatase.[13] It should also be noted that tau, APP, htt and α-synuclein have multiple phosphorylation sites on which many of the above mentioned stress kinases could act on when activated and hence they could modulate the toxic aggregation process. Therefore HSPs have the potential to indirectly impact on the assembly behaviour of each polypeptide by modulating their respective kinase activities. Most significantly, the physical state of oligomers may decide which and to what extent MAPK pathways are engaged[37] If specific HSPs regulate distinct populations of aggregation intermediates they could also impact on the balance of death/survival MAPK pathways. As molecular chaperones, HSPs may further impact in as yet unidentified ways on the various disease specific pathways implicated in AD (wnt and cdk5/p35 pathway or protein kinase C and IGF-1/insulin receptor pathway), PD (phosphatidyl inositol lipid signaling pathways) and HD (cAMP signalling cascades, mTOR signalling).

The Recruitment Hypothesis: Heat Shock Proteins as Molecular Players

Misfolding and the subsequent assembly into oligomers and higher order structures of tau, Aβ, htt and α-synuclein may result in abnormal binding and recruitment of cellular factors. The recruitment of vital proteins to deposits and their precursors may result in inactivation and/or loss of function of the molecules, abnormal intracellular localisation and abnormal activation of signaling complexes. In HD (and other polyQ diseases) it is well documented that proteins as diverse as transcription factors (e.g., CREB-binding protein/CBP, SP1, p53), HSPs and components of the UPP are recruited into aggregates.[13] Recruitment of proteasome subunits to NFTs in dystrophic neurites and LBs is also observed in AD and PD.[38] However, it should be noted that UPP proteins (and HSPs) are only found in a subset of protein deposits in some neurons. Donaldson and colleagues have also reported that proteins containing ubiquitin-binding motifs are recruited into polyQ aggregates.[39] Hence the depletion of normal ubiquitin-binding proteins caused by sequestration into aggregates may be a common event contributing to neurodegenerative diseases. The p62/sequestosome is found in tau assemblies in AD and LBs in PD supporting this hypothesis. Caspases are a further group of proteins that are recruited to aggregates: Caspase-8 is recruited to htt IBs and APP/Aβ multimers with subsequent activation probably due to the close proximity of several pro-caspase-8 molecules.[13] Hence there is evidence that recruitment of cellular proteins in these diseases may play a role in neuropathology.

The recruitment/sequestration model is particularly attractive in the case of polyQ diseases such as HD where aggregates often reside in the nucleus and coaggregate with transcription factors. Aggregates formed by mutant htt also sequester normal htt that may contribute to a 'secondary' loss-of-function. Furthermore, Wetzel and colleagues proposed a structural, molecular mechanism that endows the recruitment process with some degree of specificity by reporting an inverse relationship between the size and recruitment activity of different forms of aggregates.[40] This is in line with the idea that smaller aggregates than IBs are a more toxic species. Here, HSPs could interact with misfolded proteins and aggregated species reducing the probability of abnormal associations with other cellular proteins. Indeed, Schaffar and colleagues showed that mutant htt fragments structurally destabilise the transcription factor TATA-binding protein (TBP), but HSP40/70 chaperones inhibited its deactivation by interfering with the conformational change in mutant htt (Fig. 4).[16] Whether HSPs are able to prevent yet other events as the ones described above remains to be tested.

Heat Shock Proteins and the Formation of Ion Channels

Over ten years ago "the channel hypothesis" of amyloid toxicity was proposed. This postulated that the toxic aggregated species form nonspecific porelike channels in the membranes of cells exposed to aggregating proteins.[8] Porelike activity leading to an imbalance of ion homeostasis has been suggested to occur in AD, PD and HD. α-synuclein protofibrils exist as individual spheres, elongated chains, or rings. In contrast to the monomer or fibril, α-synuclein protofibrils bind and permeabilize lipid vesicles and maybe other membranes, perhaps via a porelike mechanism similar to beta-sheet-rich, membrane-permeabilising toxins.[8] The A53T and A30P mutants of α-synuclein promote the formation of such "amyloid pores". [8] Annular, porelike protofibrils may also be responsible for Aβ permeabilizing or channel-like activity in AD: a ring-like structure has been found in in vitro aggregation assays of the arctic variant (E22G) of Aβ40.[41] As mentioned above, htt fragments also produce annular structures and could possess pore-activity (Fig. 5). Alternatively, Monoi and colleagues proposed that polyQ sequences may form a novel structure called "μ-helix" where L-amino acid polypeptide chains form cylindrical pores with a 3.7 Å diameter, large enough to accommodate passage of small ions and water.[42] Significantly, such channels form in artificial planar lipid bilayer membranes with 40-glutamine residues (the pathological threshold for many polyQ diseases including HD), but not those with 29 residues.

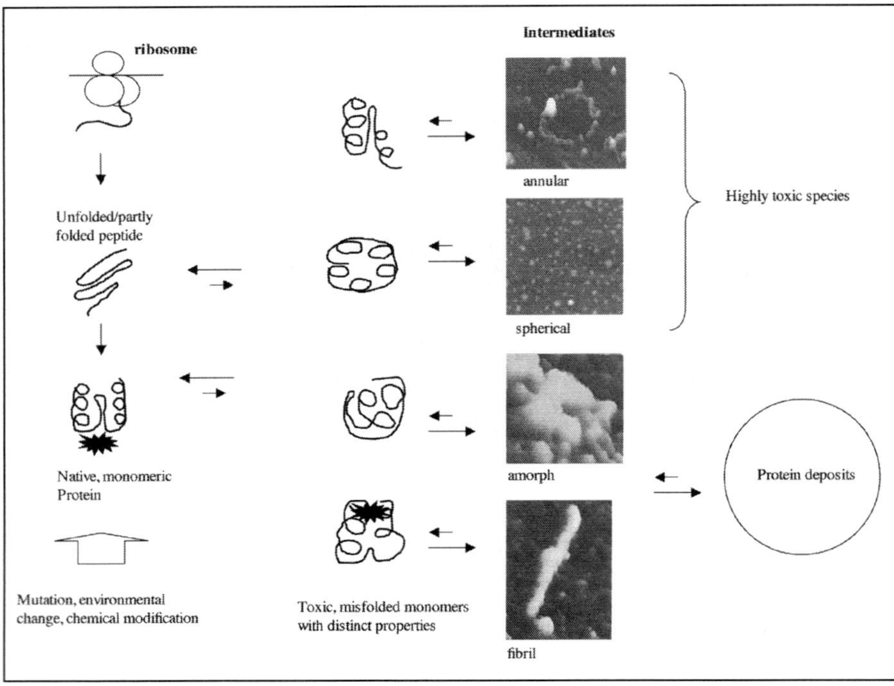

Figure 5. A disease/other protein may not correctly fold because it looses or is unable to attain its native, closely packed three-dimensional structure. Thus the peptide populates unfolded, partially folded or non-correctly folded states. In these non-native states, the peptide becomes loosely packed and hydrophobic cores become exposed to solvents enhancing the tendency to nucleate intitial oligomeric assemblies ("seeds") where the content of secondary beta structure is generally increased. The aggregation then proceeds via a multi-step process that involves intermediate species that are thought to be less stable, but more toxic than the mature fibril or final protein deposits. A monomer might misfold into various distinct conformations each of which could then give rise to a distinct higher-order assembly (spherical, annular or amorphous aggregates or a fibril). Because protein aggregation may be a generic property of polypeptide chains that does not depend on specific amino acid sequence and the toxic effects of pre-fibrillar organisation results from common structural features rather that from specific sequences of side chains it is likely that the formation of toxic oligomeric structures formed by different proteins show structural similarities and these have indeed been observed (for example, Aβ, synuclein and a fragment of huntingtin form annular oligomers). Wacker and Muchowski [22] hypothesized that HSPs (depicted as an "explosion") may stabilise a native or misfolded monomeric conformation and thereby not only prevent the intra-molecular transition that produces oligomers, but also promote a conformation that favors the production of less toxic protein deposits (e.g., inclusion bodies). Modified from reference 22. We are grateful to P. Muchowski for providing the AFM pictures.

Not only α-synuclein (see above), but also Aβ and htt bind to lipids and interactions of these hydrophobic misfolded proteins (including prion peptides) with membranes may not only cause direct damage, but also affect their aggregation behaviour frequently increasing the aggregation potential.[43,44] HSPs prevent misfolding and the exposure of hydrophobic regions and therefore may suppress the interactions of the peptides with cellular membranes. This could reduce the ability of lipids to further destabilise proteins and reduce unwanted protein-membrane interactions. HSPs could prevent channel formation since they seem to reduce the formation of annular, perhaps pore-forming species in favour of amorphous structures and fibrils as shown for the htt fragment (Fig. 5).

HSP Communication with Protein Degradation Systems and Processing of Disease Proteins

Under conditions of stress and toxic insults degradation pathways such as the UPP and the lysosomal system could be impaired leading to accumulation of misfolded/damaged proteins. While the controlled proteolysis of substrates by the UPP is crucial in controlling the levels of short-lived proteins, the lysosomal pathway contributes to total rates of protein degradation and organelle turnover. The lysosome is the catabolic factory in eukaryotic cells to which proteins can be transported following several different pathways including endocytosis, micro- and macroautophagy and chaperone-mediated autophagy. Evidence for a causal relationship between UPP impairment and neurodegeneration comes from studies on PD where mutations in Parkin and UCHL1 are linked to familial PD. Although LBs are associated with UPP components and both monomeric and aggregated α-synuclein selectively binds to a subunit of the 19S cap [45] suggesting a link to dysfunctional proteasomal degradation it is currently unclear whether α-synuclein is a substrate for the proteasome. Wildtype α-synuclein appears to be selectively translocated into lysosomes for degradation by chaperone-mediated autophagy and the A53T and A30P mutants may act as uptake blockers, inhibiting both their own degradation and that of other substrates.[46] UPP impairment has also been reported in a model of sporadic PD where systemic complex I inhibition with rotenone induces dopaminergic degeneration of the nigrostriatal pathway and formation of LBs (Fig. 3).

There are several studies that reported a reduction in proteasome- and ubiquitination-activity in patients with AD. Tau PHFs reduce proteasome activity in AD brain and Aβ inhibits the proteasome in neuronal cells.[38,47] Furthermore, presenilins and APP appear to be actively degraded by the proteasome under normal conditions and a subtle inhibition of proteasome activity could lead to an accumulation of PS and APP and their fragments and hence enhanced Aβ production. Activation of the neuronal lysosomal system and pathways converging to the lysosome, namely endocytosis and autophagy, is also a feature of brain pathology in AD and several studies support the view that progressive alterations of lysosomal function observed during aging and AD may impact on neuropathology.[48]

Studies on HD cellular models show both UPP abnormalities and interference with proteasome inhibitors modulate htt aggregation.[13,49,50] However, in vivo studies do not confirm these results so far. Cell-type specific analysis is needed and the ageing process needs to be taken into consideration too. Eukaryotic proteasomes cannot cleave within polyQ stretches and hence it is unknown how long perfect polyQ sequences are disposed of.[51] Autophagic degradation is an alternative route for disposal that has been implicated in HD.[52]

HSPs facilitate the ubiquitination and degradation of many proteins. Misfolded proteins such as mutant htt may be maintained in a degradation-competent state by HSPs and their cochaperones directing them more efficiently to degradation machineries. Indeed, the cochaperone CHIP (C-terminus of Hsp70 interacting protein) associates with expanded htt and its overexpression suppresses htt aggregation in an Hsc70 dependent manner.[53] Because HSPs appear to promote the formation of fibrils and amorphous aggregates (Fig. 5) they may facilitate aggresome formation and subsequent degradation via macroautophagy (Figs. 3,4). HSPs are also required for lysosomal protein degradation. Hsc70 forms a chaperone complex at the lysosomal membrane that is needed for unfolding the substrate proteins to efficiently translocate and degrade them in the lumen during chaperone-mediated autophagy. Therefore this process may critically regulate α-synuclein degradation (see above). HSP27 associates with the proteasome [54] and our results suggest that HSP27 partially restores proteasome activities impaired by mutant htt (W. Firdaus, A. Wyttenbach, P. Arrigo, unpublished observation). HSPs may also improve decreased proteasome activity in AD and PD. However, further research must determine whether HSPs beneficially interact with various signalling components that regulate autophagy or prevent deleterious processes arising from protein aggregation in each disease such as oxidative stress and hence prevent secondary damage to the UPP. Finally, since tau, APP and htt are caspase substrates (Fig. 1) and HSPs are known to prevent caspase activation, HSPs could suppress proteolysis of these proteins and diminish their aggregation potential.

Early Neuropathology, Synaptic Function and Axonal Transport

Early symptoms in AD correlate with dysfunctional cholinergic and glutamatergic synapses and alterations in synaptic efficacy prior to frank neurodegeneration may be caused by soluble, oligomeric assemblies of Aβ.[1] Early synaptic dysfunction and /or loss prior to overt cell death may also occur in PD and HD. Both wildtype htt and α-synuclein localise to and are abundant at the synapse and a role in synaptic homeostasis has been suggested for both proteins. In HD mouse models, mutant htt inhibits the uptake of glutamate in synaptic vesicles and a reduction in long term potentiation and depression and other abnormalities in the hippocampus have been reported.[55] Together with the finding that there is a loss of specific synaptic proteins and impairment of endo- and exocytosis in HD[55], the current results suggest an early deficit in the machinery that regulates neurotransmitter release. In PD, mutations in α-synuclein alter its association with vesicular membranes, disrupting normal recycling of vesicles, which in turn can decrease the capacity of vesicular storage and increase cytosolic levels of dopamine and its toxic metabolites.[8] Additionally, since both mutations in Parkin and UCH-L1 are linked to PD it is conceivable that a general impairment of the UPP in the synaptic compartment contributes to PD.

Could HSPs modulate synaptic pathology? Hsp40/70 members and sHSPs are present at the synapse and are increased after hyperthermia in animals.[56] Mild metabolic stress increases the resistance of synaptic terminals to Aβ dysfunction through increased expression of Hsp70 and GRP78.[57] Furthermore, the synaptic compartment contains a specific trimeric chaperone machinery consisting of cysteine string protein (CSP), small glutamine-rich tetratricopeptide repeat (TPR) containing proteins (SGTs) and HSC70 and it appears that these chaperones significantly maintain synaptic integrity[58] and that alterations of these molecules could induce pathology. Indeed, SGT proteins redistribute to inclusion bodies in HD mice and SGT expression levels are significantly decreased.[59] It remains to be seen if these chaperones are impaired in AD and PD.

Results obtained from in vitro and in vivo models of HD show that fast axonal transport is impaired due to a polyQ expansion.[60] Additionally, htt specifically enhances vesicular transport of BDNF along microtubules and a loss of this function may contribute to pathogenesis.[61] PD α-synuclein mutations show defective axonal transport in cultured neurons[62] and mutations in APP and tau disrupt axonal transport in cultured neurons and in animal models.[63] HSPs are expected to modulate axonal transport. Initial evidence for this idea comes from a study showing that elevated expression of Hsc70 in Drosophila improved axonal transport impaired by mutant htt.[60] A role for HSPs during abnormal axonal/dendritic transport in AD and PD has to be established. In this respect it is interesting to note that HSPs can be synthesised in axons after injury.[64] Finally, it should not be forgotten that glial cells critically impact on synapse and axon function and glial HSPs are highly inducible.

Heat Shock Proteins and Therapy

Animal studies support the view that modulating the levels of HSPs could prove benefical for the treatment of AD, PD and HD. However, an accurate balance (stoechiometry) of HSPs will be required to maximize their benefical roles. Excessive up-regulation of HSPs leads to unwanted effects such as cancer. A promising first target may be the heat shock transcription factor 1 (HSF1) which regulates HSP expression in concert with HSP90 that negatively regulates HSF-1. Drugs such as geldanamycin and radicicol that block the interaction of HSP90 with HSF-1, promoting HSF1 activation and synthesis of HSPs have been found to indeed induce HSPs and confer neuroprotection in in vitro and in vivo models of AD, PD and HD.[59,65,66] Treatments using drugs that up-regulate HSPs in combination with compounds that act as chemical chaperones increasing the stability of native proteins in vitro (e.g., glycerol, trehalose, dimethyl sulphoxide) await testing and may well turn out to be a valid approach to combat not only AD, PD and HD, but also other conformational disorders such as ALS and prion disease.

Conclusions

HSPs provide a first line of defence against misfolded proteins and hence could act on early steps in misfolding disease pathology. Early synaptic and axonal abnormalities in AD, PD and HD may therefore be partially reversed by HSPs. Evidence is emerging that HSPs interact with the most toxic oligomeric precursors of protein deposits and perhaps also with misfolded conformations of monomers. This points to the ability of HSPs to prevent aberrant protein interactions due to altered conformations of disease proteins with other cellular components. But other activities of HSPs including the regulation of protein degradation, the control of signalling pathways and the redox state of the cell will be equally important to consider. Since the mechanisms by which HSPs improve neuroprotection are poorly understood, a lot more research is needed to get a glimpse of the full therapeutic potential of HSPs during neurodegeneration.

Acknowledgements

Andreas Wyttenbach is grateful to the Medical Research Council (MRC, UK) and the HighQ Foundation for financial support. André Patrick Arrigo is supported by ARC grant 4602 and the Region Rhone-Alpes. We apologize to the authors of many studies that are relevant to this chapter that could not be cited due to space limitations.

References

1. Selkoe DJ. Cell biology of protein misfolding: The examples of Alzheimer's and Parkinson's diseases. Nat Cell Biol 2004; 6(11):1054-61.
2. Hartle FU. Molecular chaperones in cellular protein folding. Nature 1996; 381:571-579.
3. Walsh DM, Selkoe DJ. Oligomers on the brain: The emerging role of soluble protein aggregates in neurodegeneration. Protein and Pept Lett 2004; 11(3):1-16.
4. Goedert M. Tau protein and neurodegeneration. Semin Cell Dev Biol 2004; 15(1):45-49.
5. Drewes G. MARKing tau for tangles and toxicity. Trends Biochem Sci 2004; 29(10):548.
6. Lee MY, Giasson BI, Trojanowski JQ. More than just two peas in a pod: Common amyloidogenic properties of tau and alpha-synuclein in neurodegenerative diseases. Trends Neurosci 2004; 27(3):129-133.
7. Magrane J, Smith RC, Walsh K et al. Heat shock protein 70 participates in the neuroprotective response to intracellularly expressed beta-amyloid in neurons. J Neurosci 2004; 24(7):1700-1706.
8. Rochet JC, Outeiro TF, Conway KA et al. Interactions among alpha-synuclein, dopamine, and biomembranes. J Mol Neurosci 2004; 23:23-33.
9. Healy DG, Abou-Sleiman PM, Wood NW. PINK, PANK, or PARK? A clinicians's guide to familial parkinsonism. The Lancet 2004; 3:652-662.
10. Zoghbi HY, Orr HT. Glutamine repeats and neurodegeneration. Ann Rev Neurosci 2000; 23:217-247.
11. Davies SW, Turmaine M, Cozens BA et al. Formation of neuronal intranuclear inclusions underlies the neurological dysfunction in mice transgenic for the HD mutation. Cell 1997; 90(3):537-548.
12. Perutz MF. Glutamine repeats and neurodegenerative diseases: Molecular aspects. Trends Biochem Sci 1999; 24(2):58-63.
13. Wyttenbach A. Role of heat shock proteins during polyglutamine neurodegeneration: Mechanisms and hypothesis. J Mol Neurosci 2004; 23(1-2):69-96.
14. Shimura H, Miura-Shimura Y, Kosik K. Binding of tau to heat shock protein 27 leads to decreased concentration of hyperphosphorylated tau and enhanced cell survival. J Biol Chem 2004; 279(17):17957-17962.
15. Dedmon MM, Christodoulou J, Wilson MR et al. Heat shock protein 70 inhibits alpha-synuclein fibril formation via preferential binding to prefibrillar species. J Biol Chem 2005; 280(15):14733-14740.
16. Schaffar G, Breuer P, Boteva R et al. Cellular toxicity of polyglutamine expansion proteins: Mechanism of transcription factor deactivation. Mol Cell 2004; 15(1):95-105.
17. Dou F, Netzer WJ, Tanemura K et al. Chaperones increase association of tau protein with microtubules. Proc Natl Acad Sci USA 2003; 100(2):721-726.

18. Yang Y, Turner RS, Gaut JR. The chaperone BiP/GRP78 binds to amyloid precursor protein and decreases Abeta40 and Abeta42 secretion. J Biol Chem 1998; 273:25552-25555.
19. Furlong RA, Narain Y, Rankin J et al. Alpha-synuclein overexpression promotes aggregation of mutant huntingtin. Biochem J 2000; 346(3):577-581.
20. Charles V, Mezey E, Reddy PH et al. Alpha-synuclein immunoreactivity of huntingtin polyglutamine aggregates in striatum and cortex of Huntington's disease patients and transgenic mouse models. Neurosci Lett 2000; 289(1):29-32.
21. Kim S, Nollen EA, Kitagawa K et al. Polyglutamine protein aggregates are dynamic. Nat Cell Biol 2002; 4(10):826-831.
22. Muchowski PJ, Wacker JL. Modulation of neurodegeneration by molecular chaperones. Nat Rev Neurosci 2005; 6:11-22.
23. Andersen JK. Oxidative stress in neurodegeneration: Cause or consequence? Nat Rev Neurosci 2004; 10:S18-25.
24. Browne SE, Ferrante RJ, Beal MF. Oxidative stress in Huntington's disease. Brain Pathol 1999; 9:147-163.
25. Tabrizi SJ, Workman J, Hart PE et al. Mitochondrial dysfunction and free radical damage in the Huntington R6/2 transgenic mouse. Ann Neurol 2000; 47(1):80-86.
26. Wyttenbach A, Sauvageot O, Carmichael J et al. Heat shock protein 27 prevents cellular polyglutamine toxicity and suppresses the increase of reactive oxygen species caused by huntingtin. Hum Mol Genet 2002; 11(9):1137-1151.
27. Mattson MP. Metal-catalyzed disruption of membrane protein and lipid signaling in the pathogenesis of neurodegenerative disorders. Ann NY Acad Sci 2004; 1012:37-50.
28. Tabner BJ, Turnbull S, El-Agnaf OM et al. Formation of hydrogen peroxide and hydroxyl radicals from A(beta) and alpha-synuclein as a possible mechanism of cell death in Alzheimer's disease and Parkinson's disease. Free Radic Biol Med 2002; 32(11):1076-1083.
29. Arrigo AP, Virot S, Chaufour S et al. Hsp27 consolidates intracellular redox homeostasis by upholding glutathione in its reduced form and by decreasing iron intracellular levels. Antioxid Redox Sigal 2005; 7(3-4):414-422.
30. Tatton W, Chalmers-Redman R, Brown D et al. Apoptosis in Parkinson's Disease: Signals for Neruonal Degradation. Ann Neurol 2003; 53(3):61-72.
31. Takuma K, Yan SS, Stern DM et al. Mitochondrial dysfunction, endoplasmic reticulum stress, and apoptosis in Alzheimer's disease. J Pharmacol Sci 2005; 97(3):312-316.
32. Hickey MA, Chesselet MF. Apoptosis in Huntington's disease. Prog Neuropsychopharmacol Biol Psychiatry 2003; 27(2):255-265.
33. Beere HM. Stressed to death: Regulation of apoptotic signaling pathways by the heat shock proteins. Sci STKE 2001; 93:RE1.
34. Kadowaki H, Nishitoh H, Ichijo H. Survival and apoptosis in ER stress: The role of protein kinases. J Chem Neuroanat 2004; 28(1-2):93-100.
35. Kouroku Y, Fujita E, Jimbo A et al. Polyglutamine aggregates stimulate ER stress signals and caspase-12 activation. Hum Mol Genet 2002; 11(13):1505-1515.
36. Harper SJ, Wilkie N. MAPKs: New targets for neurodegeneration. Expert Opin Ther Targets 2003; 7(2):187-200.
37. Bell KA, O'Riordan KJ, Sweatt JD et al. MAPK recruitment by beta-amyloid in organotypic hippocampal slice cultures depends on physical state and exposure time. J Neurochem 2004; 91:349-361.
38. Ross CA, Pickart CM. The ubiquitin-proteasome pathway in Parkinson's disease and other neurodegenerative diseases. Trends Cell Biol 2004; 14(12):703-711.
39. Donaldson KM, Li W, Ching KA et al. Ubiquitin-mediated sequestration of normal cellular proteins into polyglutamine aggregates. Proc Natl Acad Sci USA 2003; 100(15):8892-8897.
40. Chen S, Berthelier V, Yang W et al. Polyglutamine aggregation behavior in vitro supports a recruitment mechanism of cytotoxicity. J Mol Biol 2001; 311:173-182.
41. Lashuel HA, Hartley D, Petre BM et al. Mixtures of wild-type and a pathogenic (E22G) form of Ab40 in vitro accumulate protofibrils, including amyloid pores. J Mol Biol 2003; 332:795-808.
42. Monoi H, Futaki S, Kugimiya S et al. Poly-L-glutamine forms cation channels: Relevance to the pathogenesis of the polyglutamine diseases. Biophys J 2000; 78(6):2892-2899.
43. Kourie JI, Henry CL. Ion channel formation and membrane-linked pathologies of misfolded hydrophobic proteins: The role of dangerous unchaperoned molecules. Clin Exp Pharmacol Physiol 2002; 29:741-753.
44. Kegel KB, Sapp E, Yoder JL et al. Huntingtin associates with membranes by directly binding phospholipids. Washington, DC: Society for Neuroscience, 2004:445.17, (Abstract).

45. Snyder H, Mensah K, Theisler C et al. Aggregated and monomeric alpha-synuclein bind to the S6' proteasomal protein and inhibit proteasomal function. J Biol Chem 2003; 278(14):11753-9.
46. Cuervo AM, Stefanis L, Fredenburg R et al. Impaired degradation of mutant alpha-synuclein by chaperone-mediated autophagy. Science 2004; 305(5688):1292-5.
47. Keck S, Nitsch R, Grune T et al. Proteasome inhibition by paired helical filament-tau in brains of patients with Alzheimer's disease. J Neurochem 2003; 85(1):115-122.
48. Nixon RA, Cataldo AM, Mathews PM. The endosomal-lysosomal system of neurons in Alzheimer's disease pathogenesis: A review. Neurochem Res 2000; 25(9-10):1161-72.
49. Wyttenbach A, Carmichael J, Swartz J et al. Effects of heat shock, heat shock protein 40 (HDJ-2), and proteasome inhibition on protein aggregation in cellular models of Huntington's disease. Proc Natl Acad Sci USA 2000; 97(6):2898-2903.
50. Bence NF, Sampat RM, Kopito RR. Impairment of the ubiquitin-proteasome system by protein aggregation. Science 2000; 292(5521):1552-2.
51. Venkatraman P, Wetzel R, Tanaka M et al. Eukaryotic proteasomes cannot digest polyglutamine sequences and release them during degradation of polyglutamine-containing proteins. Mol Cell 2004; 14(1):95-104.
52. Ravikumar B, Vacher C, Berger Z et al. Inhibition of mTOR induces autophagy and reduces toxicity of polyglutamine expansions in fly and mouse models of Huntington disease. Nat Genet 2004; 36(6):585-595.
53. Jana NR, Dikshit P, Goswami A et al. Cochaperone CHIP associates with expanded polyglutamine protein and promotes their degradation by proteasomes. J Biol Chem 2005; 280(12):11635-40.
54. Parcellier A, Schmitt E, Gurbuxani S et al. HSP27 is a ubiquitin-binding protein involved in I-kappaBalpha proteasomal degradation. Mol Cell Biol 2003; 23(16):5790-5802.
55. Li JY, Plomann M, PB. Huntington's disease: A synaptopathy? Trends Mol Med 2003; 9(10):414-420.
56. Bechtold DA, Rush SJ, Brown IR. Localisation of the heat-shock protein Hsp70 to the synapse following hyperthermic stress in the brain. J Neurochem 2000; 74:641-646.
57. Guo ZH, Mattson MP. In Vivo 2-Deoxyglucose Administration Preserves Glucose and Glutamate Transport and Mitochondrial Function in Cortical Synaptic Terminals after Exposure to Amyloid. Exp Neurol 2000; 166:173-179.
58. Tobaben S, Thakur P, Fernandez-Chacon R et al. A trimeric protein complex functions as a synaptic chaperone machine. Neuron 2001; 31:987-999.
59. Hay DG, Sathasivam K, Tobaben S et al. Progressive decrease in chaperone protein levels in a mouse model of Huntington's disease and induction of stress proteins as a therapeutic approach. Hum Mol Genet 2004; 13(13):1389-405.
60. Gunawardena S, LSG. Polyglutamine diseases and transport problems: Deadly traffic jams on neuronal highways. Arch Neurol 2005; 62(1):46-51.
61. Gauthier LR, Charrin BC, Borrell-Pages M et al. Huntingtin controls neurotrophic support and survival of neurons by enhancing BDNF vesicular transport along microtubules. Cell 2004; 118(1):127-138.
62. Saha AR, Hill J, Utton MA et al. Parkinson's disease alpha-synuclein mutations exhibit defective axonal transport in cultured neurons. J Cell Sci 2004; 117(7):1017-24.
63. Terwel D, Dewachter I, Van Leuven F. Axonal transport, tau protein, and neurodegeneration in Alzheimer's disease. Neuromoleculer Med 2002; 2(2):151-165.
64. Willis D, Li KW, Zheng JQ et al. Differential transport and local translation of cytoskeletal, injury-response, and neurodegeneration protein mRNAs in axons. J Neurosci 2005; 25(4):778-791.
65. Auluck PK, Meulener MC, Bonini NM. Mechanisms of suppression of alpha-synuclein neurotoxicity by Geldanamycin in Drosophila. J Biol Chem 2005; 280(4):2873-8.
66. Klettner A. The induction of heat shock proteins as a potential strategy to treat neurodegenerative disorders. Drug News Perspect 2004; 17(5):299-306.

Chapter 8

Heat Shock Proteins in Multiple Sclerosis

Celia F. Brosnan,* Luca Battistini and Krzysztof Selmaj

Abstract

In this review, we have addressed the possible contribution of heat shock proteins (HSP) to the pathogenesis of multiple sclerosis (MS), a chronic inflammatory demyelinating disease of the central nervous system (CNS). A particular focus of the review is on the families of HSP27, HSP60 and HSP70 because there is good evidence for both RNA and protein that expression levels of these HSP are altered in lesioned areas of the CNS. Using a variety of different approaches, the data support a role for these HSP in the generation of the immune response, particularly in the more chronic phases of the disease process. In addition, we review evidence supporting a protective role for these HSP in the injured CNS. This dual role of HSP makes an analysis of their effects in degenerative CNS diseases difficult to determine with certainty. Nevertheless, ongoing data are persuasive that this remains an important area of research that is likely to continue to contribute to our understanding of disease pathogenesis in MS.

Introduction

Multiple sclerosis is a degenerative condition of the central nervous system (CNS) that affects approximately 10^6 persons worldwide. Disease onset is usually first noted in young adulthood and epidemiological studies have shown that people in higher latitudes are more frequently affected. In the early stages, disease activity usually displays a relapsing-remitting course, but with time patients enter a secondary progressive phase with increasing evidence of disability. Clinical symptoms depend upon the location of pathological lesions within the CNS and can vary significantly. The clinical diagnosis of MS requires two features: dissemination of the disease in space and in time. The first criterion can be achieved by demonstration on MRI examination of disseminated lesions throughout the CNS, whereas the second criterion requires either occurrence of a second clinical episode or a new lesion on MRI image that presents after the onset of the initial symptoms. In the CNS, lesions are predominantly located in white matter tracts, particularly at periventricular sites. Acute active lesions are characterized by loss of myelin in the presence of a mononuclear inflammatory infiltrate centered around vessels consisting of activated lymphocytes and macrophages. Elevated levels of immunoglobulins and complement components have also been detected in the lesion as well as in cerebrospinal fluid (CSF). Older lesions show less evidence of ongoing inflammatory activity, and are characterized by demyelinated axons embedded in an astrocytic scar, reduced numbers of oligodendrocytes, and variable axonal loss.[1-3]

The presence of activated lymphocytes and immunoglobulins within the lesions and in the CSF have suggested that MS may have an immune component, and animal models have demonstrated that sensitization against CNS tissues leads to an inflammatory demyelinating

*Corresponding Author: Celia F. Brosnan—Albert Einstein College of Medicine, Department of Pathology, 1300 Morris Park Avenue, Bronx, New York 10461, U.S.A. Email: Brosnan@aecom.yu.edu

Heat Shock Proteins in Neural Cells, edited by Christiane Richter-Landsberg.

disease that displays many of the clinical and pathological characteristics of MS.[4,5] Indeed, it is now well accepted that antigens associated with the myelin sheath or with the myelin forming cell the oligodendrocyte, such as myelin basic protein (MBP), proteolipid protein (PLP) and myelin oligodendrocyte glycoprotein (MOG) are potent antigens, eliciting both T and B cell responses that, to varying degrees, can either alone or in concert passively transfer disease activity to naïve animals. Using well-defined antigens, or specific peptides derived from these antigens, further studies in animals have indicated that CD4+ T cells displaying a Th1-like cytokine profile are critical for disease expression, whereas CD4+ T cells expressing Th2-type cytokines may be protective. Of particular note for the subject matter of this chapter is the observation that animals that display a relapsing-remitting course show a switch in the predominant myelin antigen to which they respond as the disease progresses, a phenomenon known as epitope spreading, and for which there is now compelling evidence that this reflects de novo sensitization to myelin and possibly nonmyelin antigens within the inflamed CNS.[6-8]

These data have led to the concept that MS may represent an autoimmune disease mediated by T cells directed against myelin antigens, perhaps exacerbated by the presence of antibodies directed against the same or even different myelin-specific antigens. However, subsequent studies showed that myelin autoreactive T cells can be found in the circulation of most healthy individuals, raising the possibility that in MS it is a failure of normal immunoregulatory circuits that keep autoreactive T and B cells in check, which leads to disease expression.[3,9,10]

Autoimmune responses could, therefore, account for the myelin and oligodendrocyte loss, but at the present time autoimmune responses have not been implicated in initiating the axonal loss. Rather, this is thought to reflect the severity of the inflammatory response and the formation or release of toxic factors, such as peroxynitrites and glutamate, within the lesion.

The question that arises in the context of this current volume is how might heat shock proteins (HSP) contribute to the pathogenesis of MS? The term HSP covers a number of different families of proteins that are generally classified according to their molecular weight. Although historically defined by their upregulation in response to heat or stress, the constitutive expression of some HSP during embryogenesis and in the adult suggest a role in normal development and differentiation. HSP participate in protein synthesis and in the organization, structural stability, and anchoring of the cytoskeleton in both stressed and unstressed cells. These associations, mediating correct assembly and stabilizing proteins at times of vulnerability to denaturation, while not becoming integral components of the mature assembled structures, has earned them the designation of molecular chaperones or chaperonins.

Certain HSP, particularly HSP60 and HSP70, have been found to elicit strong immunological reactions.[11] In many bacterial and parasitic infections, these proteins are significantly upregulated following invasion of the host, and immune responses against them are common. The fact that these HSP of bacterial and parasitic origin share significant sequence homology with mammalian HSP has suggested that the expression of these proteins on stressed host cells may contribute to the development of autoimmunity.[8,12] In this review, we will focus on the data that support a potential role for HSP in the immune and inflammatory responses that contribute to MS and/or its animal model experimental autoimmune encephalomyelitis (EAE). Most of these data have been collected for members of the 27 kD/α-B crystallin, 60 kD and 70 kD families, and this review will be focused on these HSP. This is not to say that other HSP such as gp96 are not involved in MS or its animal model EAE,[13] but that at the present time there is less information that supports such an involvement.

The Small Heat Shock Proteins

In mammalian cells the small heat shock proteins represent a relatively small family of proteins with molecular weights of 20-28 kD in mice and 20-28 kD in humans.[14,15] The αB crystallin of the lens has been shown to share significant sequence homology with HSP27 and is considered a member of this family. The small HSP are expressed constitutively in a predominantly nonphosphorylated form in many cell types and as a multimeric 200 to 800 kD complex found in the perinuclear region of the cell. Following heat stress, the genes are induced and large protein aggregates of 800 to 2000 kD MW form that translocate to the nucleus, as well as to the cytoskeleton. Following stress, small HSPs are also rapidly mono- or diphosphorylated by specific serine protein kinases, with downregulation occurring by specific protein phosphatases.[15,16]

Interest in a role for the small HSP, and in particular αB crystallin, in MS pathogenesis first surfaced in 1995 with the observation by van Noort and colleagues that myelin isolated from MS tissues, but not from normal brains, elicited responses in T cells from both MS patients and controls.[17] The most highly immunogenic protein associated with this myelin preparation obtained from MS lesions was identified as αB-crystallin. Immunohistochemical examination of MS lesions then revealed the presence of elevated expression of αB-crystallin in both oligodendrocytes and astrocytes in association with the lesion, particularly at the lesion edge, that was not found in comparable cell populations in normal appearing white matter. In early lesions, intense immunoreactivity was predominantly localized to oligodendrocytes whereas in older lesions, reactive astrocytes were the most prominently stained. The authors concluded that αB-crystallin represents an immunodominant antigen for human T cells when expressed at the elevated levels found in active MS lesions.[17]

Since that original observation, several publications from both this same group of investigators, as well as from others, have provided strong additional support for a role for αB crystallin in the pathogenesis of MS.[18-24] For studies that have directly examined MS tissues, perhaps the most compelling data come from analyses of both mRNA and protein expression using a variety of techniques. Using semi-quantitative immunoblotting techniques and antibodies that recognize HSP27 and αB-crystallin, levels of HSP27 were found to be increased over that detected in normal controls, with the greatest increases found in active plaques. In contrast, in tissues extracted from normal appearing white matter adjacent to the MS lesions, levels were comparable to that found in normal brain tissues.[25,26] For these lesions, immunoreactivity was predominantly localized to reactive astrocytes, endothelial cells and axons at the lesion edge. In active lesions, hyperplastic interfascicular oligodendrocytes at the lesion edge were also immunoreactive. Perhaps of even greater interest was the observation that pronounced expression of HSP27 was also found in myelin isolated from MS plaques but not in myelin isolated for normal control tissue—strongly supporting the original data of van Noort and colleagues.[27] Thus, white matter undergoing immune-mediated destruction in MS lesions has been associated with altered distribution and expression of HSP27. A similar increased immunoreactivity for αB-crystallin has also been noted in Alzheimer tissue, as well as other neurodegenerative diseases, suggesting that this effect is not specific for MS but can also be found in other neurodegenerative conditions.[15]

The special sensitivity of oligodendrocytes to stress situations and an associated upregulation of αB-crystallin has gained further support from recent in vitro studies using inhibitors of the ubiquitin pathways.[28] These data show that in cultured oligodendrocytes proteasomal inhibition leads to the ubiquitination of tau and its association with αB crystallin, forming aggregates in the cell cytoplasm. As such, these aggregates resemble those found in oligodendrocytes in a number of neurodegenerative diseases of the CNS, as detailed elsewhere in this volume

Support for increased expression of small HSP in MS has also come from microarray analysis of RNA extracted from active MS lesions. In these lesions, αB-crystallin was found to be the gene with the highest (approximately 10-fold) transcriptional upregulation in MS brains when compared to normal brain tissues.[29] In our own studies, we have found a similar upregulation of αB-crystallin in RNA samples derived from the edge of chronic-active lesions (unpublished observations).

The precise role that small HSP may play in disease pathogenesis, however, remains unclear. Several attempts have been made to induce an EAE-like disease in rodents using αB-crystallin as antigen. These studies have underscored the immunogenicity of αB-crystallin for T cells, and have defined the immunodominant epitopes for Lewis rats, SJL mice and Biozzi ABH mice.[30] However, none of these epitopes induced an encephalitogenic response in naïve animals, except for the cryptic epitope 1-16 in Biozzi mice. The authors ascribe this result to the normal tolerizing mechanisms that prevent the induction of EAE by the widely expressed αB-crystallin in normal tissues, and suggest that in contrast the cryptic epitope can circumvent these mechanisms and contribute to pathogenic myelin-directed autoimmunity following T cell activation.[30] It is important to note, however, that in mice sensitized with myelin oligodendrocyte glycoprotein (MOG) the authors detected a strong T cell response to αB crystallin during the chronic phase of MOG-induced EAE. Since these responses were not detected during the acute phase of the disease, the data support the conclusion that T cell sensitization to this protein develops as myelin damage accumulates. Thus, whereas small HSP may not be able to directly elicit an encephalitogenic response, the data strongly support the conclusion that with time and exposure to damaged myelin these responses do develop and could contribute to the more chronic, destructive phases of the disease.[30] This is difficult to demonstrate in vivo but perhaps specific tolerogenic interventions, that have been used to document other instances of determinant spreading could be used to substantiate this possibility.[31]

In an interesting extension of this work, studies by Starckx and colleagues have found that αB-crystallin is a substrate for gelatinase B (matrix metalloprotease -9).[32] These authors showed that digestion of αB-crystallin with this enzyme generated the immunodominant epitopes, as well as cryptic epitopes, known to elicit T cell responses in rodents.[30] Since MMP9 is known to be potently upregulated at sites of inflammation, including in MS lesions,[33,34] this interaction could lead to the generation of epitopes capable of activating T cells present within demyelinating and inflammatory MS lesions.

The HSP60 Family

Members of this family are widely distributed in nature and homologues have been identified in most living organisms.[8,12,35] Antibodies that are unique for the bacterial and mammalian forms have been developed, but cross-reactive epitopes have also been observed. In mammalian cells, constitutive expression of HSP60 is predominantly located to mitochondria, where it plays an essential role in the folding and assembly of transported proteins into the mitochondrion in an ATP-dependent manner.[36] Following heat shock or exposure to stress HSP60 is significantly upregulated and may show altered distribution in the cell, including expression on the cell membrane.[37]

A potential role for immune responses to HSP60 in inflammatory demyelinating diseases of the CNS was first proposed by Mor and Cohen.[38] In an analysis of antigen responses of T cells isolated from the CNS of animals with EAE they noted that many of these T cells responded to HSP60. Of particular note was the observation that even in animals sensitized by the passive transfer of MBP-reactive T cells, HSP60 reactive T cells could be detected in the CNS. These data strongly suggest that de novo sensitization to HSP60 occurred within the inflammatory infiltrate.

A role for T cells responding to HSP60 in disease pathogenesis has also been proposed in animal models of arthritis and diabetes.[12] In adjuvant arthritis induced by sensitization with heat killed mycobacteria, autoaggressive T cells can be shown to recognize both mycobacterial HSP65 and cartilage glycoproteins. Perhaps the most convincing data that supports a pathogenic role for these cells in disease expression come from studies using a mycobacterial HSP65 in a tolerogenic protocol, which fully protected against adjuvant arthritis.[39]

Perhaps somewhat surprisingly, however, it has historically proved difficult to demonstrate T cell responses to HSP60 in patients with MS within either the αβ or γδ T cell populations, at least when using proliferation as a read-out of T cell reactivity.[40] In contrast, other studies have documented the presence of anti-HSP60 antibodies in both the CSF and serum of affected individuals.[41,42] Antibody titers have been found to show a high degree of correlation with the extent of disability, as determined by the Kurtzke disability score, and to increase with chronicity.

Using antibodies specific for both the mammalian and mycobacterial forms of HSP60/65, several studies have investigated whether altered levels and or distribution of these proteins are evident in MS lesions. Using immunoreactivity in mitochondria as an indicator of successful and appropriate staining, these studies have shown enhanced expression in astrocytes, oligodendrocytes, microglia, endothelial cells and inflammatory cells in lesioned areas of the brain.[40] A similar distribution has been noted in EAE, with increased expression observed on glial elements in the chronic phase of the disease.[37] Semi-quantitative immmunoblotting confirmed a small but significant increase in HSP60 during the chronic phase in some of the animals. These data support the conclusion that in EAE, although HSP60 is not markedly upregulated on CNS elements during the acute phase of the disease, with a more persistent exposure to stress HSP60 can be observed on glial cells adjacent to the lesion, as well as on debris-laden macrophages, similar to what has been found in chronic-active MS lesions.[43]

As noted earlier, in addition to immunoreactivity localized to mitochondria, staining was also detected within the cytosol at inflamed sites. That at least some of this immunoreactivity represents expression at the plasma membrane has been supported by analysis of membrane preparations isolated from inflammatory infiltrates.[37] This expression of HSP60 on the surface of inflammatory cells is consistent with reports that HSP may be present on the cell membrane of stressed macrophages and virus-infected leukocytes, where it could function as a target of activated T cells.[8] This has now been convincingly demonstrated by Kaufmann and colleagues using cytotoxic activity as a read-out for T cell responses.[8,44]

This observation also raises the possibility that cell-membrane expression by immunocompetent cells could lead to down-regulation of the immune response through the action of cytotoxic T cells directed against this epitope. This possibility has now been investigated in depth by Cohen and colleagues.[45] In their early studies of regulatory pathways involved in EAE, these authors showed that vaccination using formalin-fixed encephalitogenic T cell lines or clones could protect against EAE. Although it was well established that EAE-effector T cells needed to be activated prior to transfer to initiate disease, this group of investigators showed that T cells used for vaccination against EAE also needed to be in an activated state, since resting T cells provided no protection. Dissection of the nature of the antigen(s) that elicited this response showed that it was complex. One response was directed against the idiotope of the TCR (anti-idiotypic), showing antigen-specificity for the response. But activated T cells that were not sensitized to myelin were also found to provide significant protection. This response to activated T cells was named an anti-ergotypic response. Ongoing studies by this group have found that ergotopes are diverse molecules, but include the TCR, the high-affinity IL-2 receptor CD25, and HSP60, and have been found to reside within both the αβ and γδ T cell populations.[45] The specific mechanism by which these cells serve to down-regulate ongoing immune responses remains to be fully defined, and may differ depending upon which specific T cell subset is involved, but include cytotoxic activity and release of regulatory cytokines such IL-10 and TGFβ. Regulatory T cell clones or lines can be derived that can be boosted by specific stimulation with isolated ergotypic molecules and which are capable of passively-transferring regulatory activity to naïve recipients.[45] However, these regulatory T cells are themselves subject to regulation and may become anergic or die after interaction with effector T cells. Thus, determining the appropriate conditions that result in the survival and retention of functionality of these T cells is of some considerable interest for management of diseases of possible autoimmune etiology such as MS.

HSP may also contribute to the immune/inflammatory response through activation of the innate immune response. Several studies have now documented a role for HSP60 as a ligand for toll-like receptors (TLR) 2 and 4.[46-49] These studies have shown that binding of labeled HSP60 to cells of the monocyte/macrophage series, which are the major sites of TLR expression, could be blocked by unlabeled HSP60, but not by HSP70, HSP90 or gp96, all of which share the α2-macroglobulin binding sites.[47] Further support for a role for TLR in macrophage activation by HSP60 has come from studies in C3H/HeJ mice, which carry a mutant TLR4, and which were found to be unresponsive to HSP60.[48] Although the exact nature of the interaction between TLR2 and 4 with HSP60 in this response is still under investigation, it would appear that mammalian as well as microbial HSPs can function in this regard.[50] Activation of TLRs leads to the generation of a wide range of cytokines, chemokines and other factors that are strongly proinflammatory, but which also play an important role in shaping the acquired immune response through effects on dendritic cells and in shaping the cytokine profile of antigen-reactive T cells involved in the acquired immune response. These data would fit well with the "danger hypothesis", which proposes that for effective activation of the immune response, danger signals are required that act as triggers for activation of the acquired immune response.[51] Expression of self-HSP, such as HSP60 and perhaps 70, on the cell surface could potentially function in this regard through activation of TLR.

The HSP70 Family

This family is composed of multiple members, all of which bind ATP, but which differ in their localization within the cell as well as in their antigenic properties. Some members are expressed constitutively and show little or no response to heat or stress, whereas others are barely detected in resting cells but are rapidly and markedly upregulated by heat and/or stress. The major constitutive form is now known as HSC70. It has a MW of 73 kD and is found within the cytosol and nucleus, and is only moderately upregulated by stress. The major inducible form is now called HSP70. It has a MW of 72 kD, and functions as a regulator of transcription as well as a molecular chaperone. Like HSP60, members of the HSP70 are found in most living organisms, and in bacteria are found in association with the cell membrane.

In a study of immunoreactivity to HSP70 in patients with MS, Salvetti and colleagues observed an increased response to mycobacterial HSP70, which was not observed when bacterial HSP65 was used as the antigen.[52,53] A similar increased proliferation to mycobacterial antigens was detected in lymphocytes within the CSF by Birnbaum et al, which was not found in lymphocytes isolated from matching peripheral blood samples.[54,55] In agreement with the results of Salvetti et al,[52] they also failed to detect a comparable response to mycobacterial HSP65. Further studies by Stinissen et al suggested that the increased responses to mycobacterial HSP70 might preferentially reside within the minor γδ T cell population rather than in the major lymphocyte population that expresses the αβ T cell receptor.[56] This possibility has also received independent support in our own work, where we detected an oligoclonal pattern to the TCR expressed only in the Vδ2Jδ3 subset in MS patients, which was not present in γδ T cells isolated from patients with OND or healthy controls.[57] The link with a response to HSP70 came with an analysis of TCR usage in T cell lines isolated from the peripheral blood of MS patients and patients with tuberculosis using mycobacterial HSP as antigen. The data showed that the sequence of the Vδ2Jδ3 TCR in the T cells lines responding to HSP70 was identical to that found in this same subset of γδ T cells in MS brain lesions.[57] Taken together, these data suggest that, at least for this very minor subset of T cells, that they are responding to HSP70.

Further support for HSP70 in the activation of T cells responses in MS has come from studies of its role in antigen presentation of myelin basic protein (MBP). As we have reviewed previously,[40] several members of different HSP families have been implicated in the antigen-presentation pathway, either by interacting with antigenic peptides during antigen processing, or during the assembly of the MHC-antigen complex, or by functioning as antigen-presenting molecules themselves.[58,59] HSP are known to deliver proteins to lysosomes for degradation and may thus facilitate the delivery of peptides to MHC molecules. Consistent with this hypothesis are the observations that peptides representing self-HSP70 can be found bound to MHC class II molecules during the biosynthetic pathway.[60] Similarly, with peptides that bind to class I MHC, HSP have been proposed to constitute a "relay line" by which peptides are transferred to MHC class I molecules.[58,59] As noted above, members of the HSP70 family have been particularly implicated in these events. Recently, we have been able to identify an important role for HSP70 in the processing and presentation of myelin antigens, particularly myelin basic protein (MBP), to myelin-reactive T cells. In the first set of experiments, we demonstrated that HSP70 associated with MBP and proteolipid protein (PLP) in MS brains.[61] In all samples of MS brains examined, coimmunoprecipitation studies demonstrated a direct association of MBP with HSP70 but not HSP90. In some brains associations between PLP and HSP70 were also found, but interactions between MOG and HSP70 was not detected. In striking contrast to the data obtained with the MS samples, no interaction between these myelin proteins and HSP was detected in normal brain samples. A similar data set was obtained for brain tissue isolated from mice sensitized to develop EAE when compared with tissue samples from normal control brains.[61] Specific binding of MBP to HSP70 in vitro was then documented using surface plasmon resonance spectroscopy. Of particular note was the observation that MBP complexed with HSP70 was taken up more avidly by antigen-presenting cells (APC) than MBP alone.

To follow-up on this observation, we then addressed the role of these HSP70/MBP complexes in the activation of T cells.[62] The results of these studies showed that the association between HSP70 and MBP in APC was ATP/ADP dependent, and could be detected as a complex within the endocytic pathway using confocal imaging. Surface plasmon resonance spectroscopy further showed that interactions between HSP70 and MBP specifically involved the immunodominant epitopes (85-99 and 80-99 for HLA-DR2). Using an MBP-specific T cell hybridoma, the functional significance of this interaction was assessed. The data showed that: 1) interactions between HSP70 and MBP enhanced proliferation responses over MBP alone when either full-length MBP or the elongated 80-99 peptide was used in the complex but not when MBP85-99 peptide was used, 2) the response showed MHC class II specificity, and 3) reduced levels of HSP70 within the cell also reduced T cell proliferation.[62] These data indicate that the critical interaction between HSP70 and MBP relates to processing of MBP rather than to presentation per se. Finally, the importance of this association was further demonstrated in transfection experiments using vectors encoding either MBP protein or HSP70-MBP or HSP40-MBP. The data showed that although transfections with MBP and HSP-MBP fusion proteins yielded similar kinetics, a significantly increased response to APC expressing the HSP70-MBP fusion protein but not the HSP40-MBP fusion protein or MBP alone was observed. Since HSP40 functions as a cochaperone for HSP70 and is a member of a closely related family, these data serve to underscore the specificity of the reaction with HSP70. Taken together, these data strongly suggest that aberrant self HSP expression may lead to the enhancement/modulation of autoimmune responses through MHC class II.[62]

In MS lesions, the distribution of HSP70 protein has been assessed using semi-quantitative western blotting and immunohistochemistry.[63] Surprisingly, the western blot data showed that levels of HSC70 were reduced by 30 to 50% over that found in normal brain tissue, whereas levels of HSP70 remained unchanged. However, the data also showed that HSC70 comprised almost 2% of total brain protein with levels slightly higher for grey matter ($2.1 \pm 0.54\%$) than white matter ($1.64 \pm 0.16\%$). These data suggest that HSC70 is an important component of normal brain function and that the lower levels of HSC70 detected in MS tissues might reflect the loss of brain function in diseased/degenerating tissues.

This raises an important issue. In most of this review, we have focused on the potential role of HSP as antigens that could contribute to immune-mediated diseases. However, HSP were first recognized for the protective effect that they provided against various kinds of insults. So, for example, it has been shown that HSP can protect against TNF-mediated toxicity. In WEHI cells, HSP70 has been shown to protect against cytokine-mediated toxicity by interfering with the activation of phospholipase A2.[64] Feinstein and colleagues have also shown that induction of HSP70 in astrocytes reduces activation of the inducible form of nitric oxide synthase, a cytokine responsive gene in these cells.[65] Thus, HSP70 activation may serve to limit the production of proinflammatory and neurotoxic factors in the CNS.[66,67] Consistent with this are several in vivo studies that have shown that expression of HSPs can protect against diseases that have an inflammatory or neurodegenerative pathogenesis.[68,69] Loss of HSC70 in MS lesions may, therefore, reflect loss of a protective factor against cytokine-mediated tissue damage. Nevertheless, HSP70 can be detected in MS lesions particularly in association with myelin profiles in the lesion center, and in reactive astrocytes at the border of chronic-active and chronic-silent lesions. More recently, mutations in the small heat shock proteins HSPB1 and HSPB8 have been shown to show associations with some distal motor neuropathies, indicating that these proteins likely play an important role in the normal functioning of neurons and/or axons. HSPB1, can also suppress the toxic effects of the polyglutamine protein, huntingtin, and decrease the levels of reactive oxygen species produced as part of the response to this toxic protein.[15] Taken together, these data show that a protective role for heat shock proteins cannot be ignored.

Concluding Remarks

In this review we have briefly touched upon the potential role of certain families of HSP in MS. What the specific role of each of these proteins play in disease expression is difficult to say with certainty, but both protective and pathogenic mechanisms are likely. The beneficial effects could include not only protection against various kinds of potentially toxic factors, but also mechanisms for down regulation of the immune response. Harmful effects might include the formation of additional antigenic targets that might contribute to the development of chronic disease. In our previous review of this subject we raised the question as to why, given the universality of the stress response, other neurological diseases in which altered expression of HSP have also been noted, such as Alzheimer's and Parkinson's diseases, do not show an inflammatory response against these proteins. At that time we suggested that it was possible to envision that in the context of an ongoing immune response, altered expression of HSP could have different outcomes from those that would follow altered expression in noninflammatory conditions.[40] Today, perhaps, it is easier to recognize similarities between these disease states than was recognized previously. Inflammatory processes are now considered to be part of the underlying pathogenesis of both Alzheimer's and Parkinson's diseases, and conversely, in MS axonal degeneration and gray matter loss are increasingly recognized as important correlates of chronic disease. A reexamination of the status of HSP in these different neurodegenerative diseases is, therefore, timely and this volume provides an opportunity to bring these disparate data sets together, which may hopefully provoke new and provocative ideas relevant to the role of these factors in neurodegenerative CNS disorders.

References

1. Weiner HL. Multiple sclerosis is an inflammatory T-cell-mediated autoimmune disease. Arch Neurol 2004; 61(10):1613-1615.
2. Steinman L, Martin R, Bernard C et al. Multiple sclerosis: Deeper understanding of its pathogenesis reveals new targets for therapy. Annu Rev Neurosci 2002; 25:491-505.
3. Lassmann H, Ransohoff RM. The CD4-th1 model for multiple sclerosis: A critical [Correction Of Crucial] Reappraisal. Trends Immunol 2004; 25(3):132-137.
4. Segal BM. Experimental autoimmune encephalomyelitis: Cytokines, effector t cells, and Antigen-presenting cells in a prototypical Th1-mediated autoimmune disease. Curr Allergy Asthma Rep 2003; 3(1):86-93.
5. Wekerle H, Kojima K, Lannes-Vieira J et al. Animal models. Ann Neurol 1994; 36(Suppl):S47-53.
6. Miller SD, Eagar TN. Functional role of epitope spreading in the chronic pathogenesis of autoimmune and Virus-induced demyelinating diseases. Adv Exp Med Biol 2001; 490:99-107.
7. Cross AH, Tuohy VK, Raine CS. Development of reactivity to new myelin antigens during chronic relapsing autoimmune demyelination. Cell Immunol 1993; 146(2):261-269.
8. Kaufmann SH, Schoel B, Van Embden JD et al. Heat-shock protein 60: Implications for pathogenesis of and protection against bacterial infections. Immunol Rev 1991; 121:67-90.
9. Bar-Or A, Oliveira EM, Anderson DE et al. Molecular pathogenesis of multiple sclerosis. J Neuroimmunol 1999; 100(1-2):252-259.
10. Oliveira EM, Bar-Or A, Waliszewska AI et al. CTLA-4 dysregulation in the activation of myelin basic protein reactive t cells may distinguish patients with multiple sclerosis from healthy controls. J Autoimmun 2003; 20(1):71-81.
11. Moseley P. Stress proteins and the immune response. Immunopharmacology 2000; 48(3):299-302.

12. Cohen IR. Autoimmunity to chaperonins in the pathogenesis of arthritis and diabetes. Annu Rev Immunol 1991; 9:567-589.
13. Chandawarkar RY, Wagh MS, Kovalchin JT et al. Immune modulation with high-dose heat-shock protein Gp96: Therapy of murine autoimmune diabetes and encephalomyelitis. Int Immunol 2004; 16(4):615-624.
14. Kappe G, Franck E, Verschuure P et al. The human genome encodes 10 alpha-crystallin-related small heat shock proteins: Hspb1-10. Cell Stress Chaperones. Spring 2003; 8(1):53-61.
15. Der Perng M, Quinlan RA. Neuronal diseases: Small heat shock proteins calm your nerves. Curr Biol 2004; 14(15):R625-626.
16. Ciocca DR, Oesterreich S, Chamness GC et al. Biological and clinical implications of heat shock protein 27,000 (Hsp27): A review. J Natl Cancer Inst 1993; 85(19):1558-1570.
17. Van Noort JM, Van Sechel AC, Bajramovic JJ et al. The small heat-shock protein alpha B-crystallin as candidate autoantigen in multiple sclerosis. Nature 1995; 375(6534):798-801.
18. Chou YK, Burrows GG, Latocha D et al. CD4 T-cell epitopes of human alpha B-crystallin. J Neurosci Res 2004; 75(4):516-523.
19. Holmoy T, Vartdal F. Cerebrospinal fluid t cells from multiple sclerosis patients recognize autologous epstein-barr virus-transformed b cells. J Neurovirol 2004; 10(1):52-56.
20. Tajouri L, Mellick AS, Ashton KJ et al. Quantitative and qualitative changes in gene expression patterns characterize the activity of plaques in multiple sclerosis. Brain Res Mol Brain Res 2003; 119(2):170-183.
21. Van Veen T, Van Winsen L, Crusius JB et al. [Alpha]B-crystallin genotype has impact on the multiple sclerosis phenotype. Neurology 2003; 61(9):1245-1249.
22. Vojdani A, Vojdani E, Cooper E. Antibodies to myelin basic protein, myelin oligodendrocytes peptides, alpha-beta-crystallin, lymphocyte activation and cytokine production in patients with multiple sclerosis. J Intern Med 2003; 254(4):363-374.
23. Celet B, Akman-Demir G, Serdaroglu P et al. Anti-alpha B-crystallin immunoreactivity in inflammatory nervous system diseases. J Neurol 2000; 247(12):935-939.
24. Bajramovic JJ, Plomp AC, Goes A et al. Presentation of alpha B-crystallin to t cells in active multiple sclerosis lesions: An early event following inflammatory demyelination. J Immunol 2000; 164(8):4359-4366.
25. Bajramovic JJ, Lassmann H, Van Noort JM. Expression of alpha B-crystallin in glia cells during lesional development in multiple sclerosis. J Neuroimmunol 1997; 78(1-2):143-151.
26. Van Noort JM, Van Sechel AC, Van Stipdonk MJ et al. The small heat shock protein alpha B-crystallin as key autoantigen in multiple sclerosis. Prog Brain Res 1998; 117:435-452.
27. Aquino DA, Capello E, Weisstein J et al. Multiple sclerosis: Altered expression of 70- and 27-kDa heat shock proteins in lesions and myelin. J Neuropathol Exp Neurol 1997; 56(6):664-672.
28. Richter-Landsberg C, Bauer NG. Tau-inclusion body formation in oligodendroglia: The role of stress proteins and proteasome inhibition. Int J Dev Neurosci 2004; 22(7):443-451.
29. Chabas D, Baranzini SE, Mitchell D et al. The influence of the proinflammatory cytokine, osteopontin, on autoimmune demyelinating disease. Science 2001; 294(5547):1731-1735.
30. Thoua NM, Van Noort JM, Baker D et al. Encephalitogenic and immunogenic potential of the stress protein alphaB-crystallin in biozzi ABH (H-2A(G7)) Mice. J Neuroimmunol 2000; 104(1):47-57.
31. Seamons A, Perchellet A, Goverman J. Immune tolerance to myelin proteins. Immunol Res 2003; 28(3):201-221.
32. Starckx S, Van Den Steen PE, Verbeek R et al. A novel rationale for inhibition of gelatinase B in multiple sclerosis: MMP-9 destroys alpha B-crystallin and generates A promiscuous T cell epitope. J Neuroimmunol 2003; 141(1-2):47-57.
33. Opdenakker G, Nelissen I, Van Damme J. Functional roles and therapeutic targeting of gelatinase B and chemokines in multiple sclerosis. Lancet Neurol 2003; 2(12):747-756.
34. Bever Jr CT, Rosenberg GA. Matrix metalloproteinases in multiple sclerosis: Targets of therapy or markers of injury? Neurology 1999; 53(7):1380-1381.
35. Bukau B, Horwich AL. The Hsp70 and Hsp60 chaperone machines. Cell 1998; 92(3):351-366.
36. Martin J. Molecular chaperones and mitochondrial protein folding. J Bioenerg Biomembr 1997; 29(1):35-43.
37. Gao YL, Brosnan CF, Raine CS. Experimental autoimmune encephalomyelitis. Qualitative and semiquantitative differences in heat shock protein 60 expression in the central nervous system. J Immunol 1995; 154(7):3548-3556.
38. Mor F, Cohen IR. Analysis of the autoimmune infiltrate in experimental encephalomyelitis. Isr J Med Sci 1992; 28(2):139-140.

39. Quintana FJ, Carmi P, Mor F et al. DNA fragments of the human 60-kda heat shock protein (HSP60) vaccinate against adjuvant arthritis: Identification of a regulatory HSP60 peptide. J Immunol 2003; 171(7):3533-3541.
40. Brosnan CF, Battistini L, Gao YL et al. Heat shock proteins and multiple sclerosis: A review. J Neuropathol Exp Neurol 1996; 55(4):389-402.
41. Gao YL, Raine CS, Brosnan CF. Humoral response to Hsp 65 in multiple sclerosis and other neurologic conditions. Neurology 1994; 44(5):941-946.
42. Prabhakar S, Kurien E, Gupta RS et al. Heat shock protein immunoreactivity in CSF: Correlation with oligoclonal banding and demyelinating disease. Neurology 1994; 44(9):1644-1648.
43. Raine CS. The dale E. mcfarlin memorial lecture: The immunology of the multiple sclerosis lesion. Ann Neurol 1994; 36(Suppl):S61-72.
44. Koga T, Wand-Wurttenberger A, Debruyn J et al. T cells against a bacterial heat shock protein recognize stressed macrophages. Science 1989; 245(4922):1112-1115.
45. Cohen IR, Quintana FJ, Mimran A. Tregs in T cell vaccination: Exploring the regulation of regulation. J Clin Invest 2004; 114(9):1227-1232.
46. Iwahashi M, Yamamura M, Aita T et al. Expression of toll-like receptor 2 on CD16+ blood monocytes and synovial tissue macrophages in rheumatoid arthritis. Arthritis Rheum 2004; 50(5):1457-1467.
47. Habich C, Baumgart K, Kolb H et al. The receptor for heat shock protein 60 on macrophages is saturable, specific, and distinct from receptors for other heat shock proteins. J Immunol 2002; 168(2):569-576.
48. Ohashi K, Burkart V, Flohe S et al. Cutting edge: Heat shock protein 60 is a putative endogenous ligand of the toll-like receptor-4 complex. J Immunol 2000; 164(2):558-561.
49. Vabulas RM, Ahmad-Nejad P, Ghose S et al. HSP70 as endogenous stimulus of the toll/interleukin-1 receptor signal pathway. J Biol Chem 2002; 277(17):15107-15112.
50. Vabulas RM, Wagner H, Schild H. Heat shock proteins as ligands of toll-like receptors. Curr Top Microbiol Immunol 2002; 270:169-184.
51. Matzinger P. An innate sense of danger. Ann NY Acad Sci 2002; 961:341-342.
52. Salvetti M, Ristori G, Buttinelli C et al. The immune response to mycobacterial 70-kda heat shock proteins frequently involves autoreactive T cells and is quantitatively disregulated in multiple sclerosis. J Neuroimmunol 1996; 65(2):143-153.
53. Salvetti M, Buttinelli C, Ristori G et al. Heat shock proteins as targets for gamma-delta T cells in multiple sclerosis. Ann Neurol 1992; 32(3):410-411.
54. Birnbaum G. Stress proteins: Their role in the normal central nervous system and in disease states, especially multiple sclerosis. Springer Semin Immunopathol 1995; 17(1):107-118.
55. Birnbaum G, Kotilinek L. Heat shock or stress proteins and their role as autoantigens in multiple sclerosis. Ann NY Acad Sci 1997; 835:157-167.
56. Stinissen P, Vandevyver C, Medaer R et al. Increased frequency of gamma delta T cells in cerebrospinal fluid and peripheral blood of patients with multiple sclerosis. Reactivity, Cytotoxicity, And T Cell Receptor V Gene Rearrangements. J Immunol 1995; 154(9):4883-4894.
57. Battistini L, Selmaj K, Kowal C et al. Multiple sclerosis: Limited diversity of the V delta 2-J delta 3 T-cell receptor in chronic active lesions. Ann Neurol 1995; 37(2):198-203.
58. Gullo CA, Teoh G. Heat shock proteins: To present or not, that is the question. Immunol Lett 2004; 94(1-2):1-10.
59. Robert J. Evolution of heat shock protein and immunity. Dev Comp Immunol 2003; 27(6-7):449-464.
60. Fremont DH, Hendrickson WA, Marrack P et al. Structures of an MHC class II molecule with covalently bound single peptides. Science 1996; 272(5264):1001-1004.
61. Cwiklinska H, Mycko MP, Luvsannorov O et al. Heat shock protein 70 associations with myelin basic protein and proteolipid protein in multiple sclerosis brains. Int Immunol 2003; 15(2):241-249.
62. Mycko MP, Cwiklinska H, Szymanski J et al. Inducible heat shock protein 70 promotes myelin autoantigen presentation by the HLA class II. J Immunol 2004; 172(1):202-213.
63. Aquino DA, Klipfel AA, Brosnan CF et al. The 70-kDa heat shock cognate protein (HSC70) is a major constituent of the central nervous system and is up-regulated only at the mRNA level in acute experimental autoimmune encephalomyelitis. J Neurochem 1993; 61(4):1340-1348.
64. Jaattela M. Overexpression of major heat shock protein hsp70 inhibits tumor necrosis factor-induced activation of phospholipase A2. J Immunol 1993; 151(8):4286-4294.
65. Feinstein DL, Galea E, Aquino DA et al. Heat shock protein 70 suppresses astroglial-inducible nitric-oxide synthase expression by decreasing NFkappaB activation. J Biol Chem 1996; 271(30):17724-17732.

66. Paidas CN, Mooney ML, Theodorakis NG et al. Accelerated recovery after endotoxic challenge in heat shock-pretreated mice. Am J Physiol Regul Integr Comp Physiol 2002; 282(5):R1374-1381.
67. Van Molle W, Wielockx B, Mahieu T et al. HSP70 protects against TNF-induced lethal inflammatory shock. Immunity 2002; 16(5):685-695.
68. Jacquier-Sarlin MR, Fuller K, Dinh-Xuan AT et al. Protective effects of Hsp70 in inflammation. Experientia 1994; 50(11-12):1031-1038.
69. Klettner A. The induction of heat shock proteins as a potential strategy to treat neurodegenerative disorders. Drug News Perspect 2004; 17(5):299-306.

Index

H

I

L

M

N

O

P

R

S

T

U

V